PRAIRIE EXPERIENCES

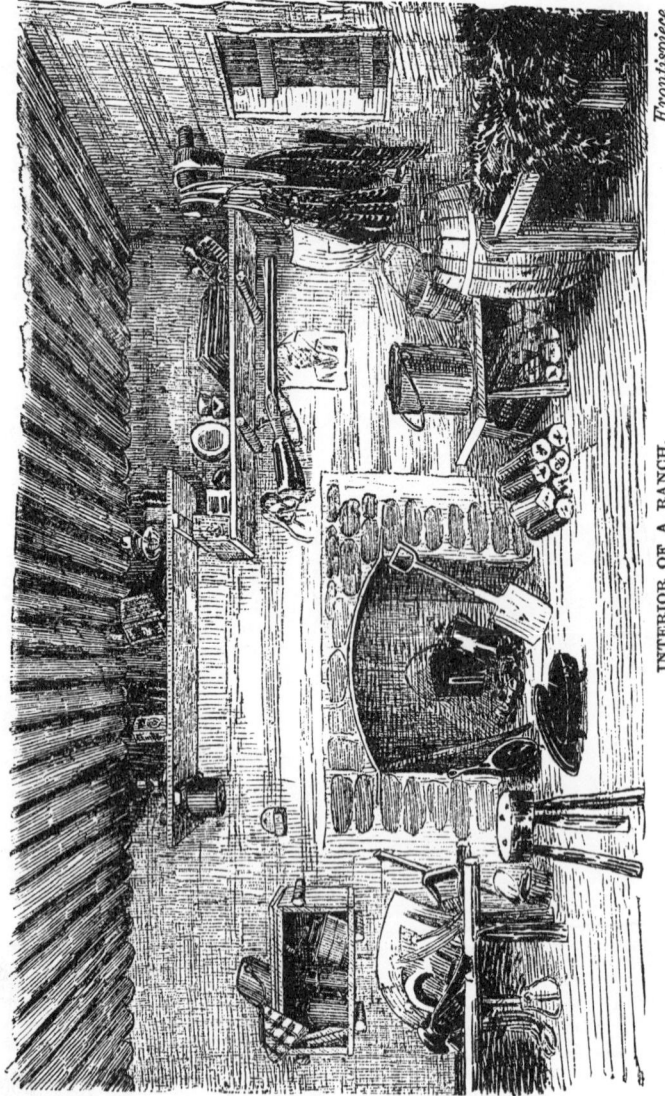

INTERIOR OF A RANCH. *Frontispiece.*

PRAIRIE EXPERIENCES

IN

HANDLING CATTLE AND SHEEP

BY
WILLIAM SHEPHERD

ILLUSTRATED

BOOKS FOR LIBRARIES PRESS
FREEPORT, NEW YORK

First Published 1885
Reprinted 1971

INTERNATIONAL STANDARD BOOK NUMBER:
0-8369-5964-7

LIBRARY OF CONGRESS CATALOG CARD NUMBER:
70-165807

PRINTED IN THE UNITED STATES OF AMERICA

LIST OF ILLUSTRATIONS.

INTERIOR OF RANCH	*Frontispiece*
	PAGE
RANCH IN WYOMING	24
LASSOING ON THE PRAIRIE	38
A SAGE HEN	132
CATTLE BRANDS	182

PRAIRIE EXPERIENCES.

To write a book after travelling in America seems inevitable. Is it a disease catching from one traveller to another? or is it caught in the country? The subject is so large, the parts so various and changing, that however much you may have studied the country in books, you cannot help seeing and hearing much that is new to you and vanity suggests new to others. We all start with a good intention of writing nothing; we break down in our resolves, and are often delivered of crude and hastily-formed notions, whose faults we do not see till others tell us. But does not every parent look on his child for a time as a novelty, a necessity, and an object of public interest. The following narrative has not been compiled from guide-books or statistical pamphlets. I have told what I saw and experienced, and have not ventured to deal with social problems or large generalizations. Before, therefore, starting on a trip through these pages, let me say that my journey took me through Wyoming, about Montana, across to Washington Territory, then back from California, through Nevada into Idaho. Cattle, cow-boys, round-ups, sheep-driving, herders, and life on the prairie, these are my text, to which I will strive to stick. As for the fifty millions of people, white, black, and yellow, who inhabit the States, twenty months spent in one corner do not privilege me to generalize. My time was spent mostly in the midst of the prairies; there, naturally, I could learn nothing of the east, centre, or south of the great Republic; nothing about her ever-growing cities, their trade and their peculiarities. Away from the haunts of men one seldom met any of the upper and educated classes, and the pleasures of social and literary intercourse are for the time superseded. The life is sometimes pleasant, sometimes dreary; there is plenty of exposure and not a little discomfort; there is generally good health, and consequently good temper; there are all sorts and conditions of men, who meet you on perfect equality, whether better or worse than yourself; your wants are few, as generally you have to satisfy them yourself. It is wonderful how you lop off necessities when they burden your time and occupations. You have entered on a new life in a new world. It is not all admirable, for good and evil are everywhere balanced. With freedom in forming new opinions, you are apt to grow disdainful of the small niceties of civilization; the trammels of society are cast off, leading to a dangerous drop into rude habits and ill-restrained language; the impossibility of fulfilling all the refinements of the toilet engender a disregard of personal neatness. Much cannot be helped; some might be avoided. Young men are naturally the more easily influenced by their surroundings, and fall too readily into the habits and tricks of speech most honored on the prairies. This is the main educational disadvantage to young men starting alone in the West, for good breeding has to be nurtured by descent and association; coarseness may be learned in a day. The traveller landed in America, however, has his mind filled with serious projects: he hopes to find something to do, something which will keep him for the present, and promise a chance of ease in the future. We have all heard of the gains in cattle-ranching—is not thirty per cent a common return? Movrick, one of the cattle-kings, began, it is pointed out, with a single steer and a branding-iron, and now his herds run on a hundred hills. There is money in that. How can we do likewise? The times for these marvels are gone by. After the war a vast number of unclaimed cattle were running loose; they were the spoil of whoever could rope them; then those who bought even ten years ago got their herds at a very low average of seven or eight dollars. If you have plenty of room and good feed, you may expect eighty or ninety per cent of calves to cows; if, at the same time, the all-round price advances to twenty-two or twenty-four dollars, the chances have been much in your favor. Now, the expenses in a crowded country of searching out your cattle at the different round-ups, and in parts of the country lying sixty miles from your range, mount up to such a degree, that with less than 5000 head the stock-owner's profit is far below the normal. Men still do start with small numbers, but they have first to seek a very secluded spot, and then must be constantly riding round and driving the cattle back on to the home range: this injures the cattle; there is great loss of labor in doing continually over again the same work, and the rancher is forced to lay in a stock of hay to feed his cattle in the winter time, as he cannot allow them to roam at liberty and fight for themselves.

For men with small capital it will prove nowadays more profitable to keep sheep; the tendency heretofore has been too exclusively to breed for fine wool, but now mutton is much in request, and a big carcase is aimed at. I do not know that any mutton is at present exported, but sheep are multiplying considerably in Colorado, Wyoming, and Montana; it will not be long before mutton will follow beef as an import to this country. Sheep are delicate animals compared with cattle and horses, and of course cannot be let run wild as the other stock is; but if a man understands them, will live on the range to superintend his herders, and has a farm where he can put up a good quantity of hay for winter use, he will get more profit out of his three to five thousand sheep than out of the same money expended on cattle. The business is not popular, and the sheep man, unless he lives quite apart, is always in discord with his neighbors who own other stock. The man who intends to farm will not probably go so far West as he who wishes to rear stock; the new farmers are settling into Dakota, Nebraska, and Kansas, where promising land can be homesteaded or pre-empted; but a very large number of new emigrants who can afford to pay something for improved estate find what they want without travelling so far.

If you wish to get any information in America you must "go and see for yourself"; this phrase is constantly the concise answer to the inquirer concerning the West. It is excellent advice; it may probably often cloak a certain amount of ignorance concerning the uttermost parts of their States which the dwellers on the Atlantic shores do not care to admit; but no man could hope to answer all the questions of strangers. Probably due to the influence of our school atlases, which gave no larger sheet to North America than it did to France or Scotland, we start with dwarfed ideas and notions confused as to distances measured in thousands of miles. It takes much travel to appreciate the proportions of the new country. On first arriving, therefore, you are apt to fancy that you can gather knowledge while staying in some of the large Eastern towns, leaning on your experience of the amiable landlords and talkative inhabitants of the Continent; disabuse yourself of the idea, or you will certainly be disappointed. To know precisely anything of a neighboring State you must go there, and travel a day to reach it. Sometimes railway or land offices pretend to supply information, but as a rule they know nothing, and conceal that they don't know. A railway official will sell you a ticket to San Francisco, New Orleans, or Quebec, to be reached in the shortest time by traversing their line to commence with, in their solid trains of palatial Pullman sleeping-cars, magnificent drawing-room and world-renowned dining-cars; but whether, if you need to change lines, you will find a correspondence of trains, or whether any object of interest along the road would justify a delay, is quite out of the official business. Buy your ticket and go about your business, and let the clerk go on with his discussion of last night's music-hall programme, or finish the newspaper's description of horrors and dog-fights, and the detailed interview of Your Special Correspondent on some intimately private family matter.

It's a new world, and a very different world, fortunately; for travellers who hope to do more than move up and down and stare at buildings, the common language is a pass-key for the Englishman, if he will make use of it. Whether he wishes or not, the new surroundings will influence him; a short month, and he ceases to be surprised; in six months he will find himself largely Americanized, and judge differently the subjects he at first felt inclined only to criticise. Second thoughts are not always best; newer judgment is influenced by later prejudices, and may be more incorrect than one's first opinions; in most cases, however, it will be more kind. The majority of Englishmen accept readily this new baptism, and fall into line with American ways of life and thought, for which they deserve some credit and regard from their cousins at least. On the contrary, Americans seldom care to conceal, and often in their conversation and newspapers explicitly declare, their dislike of their nearest relations taken generally, and nourish a hostility which desires to be fed by injuries to England and her trade. This dislike is the only sentimental side of the American's political views. It is a tradition with them that the British monarch stands for despotism and aristocratic persecution of the people; and to this idea alone would they sacrifice their positive interests in connection with foreign countries; for the rest, admit their wheat and bacon duty free, and go your ways with your old-world machinery. For oppressed subjects there is always escape to free America. France having turned republican gratifies the vanity of the citizens by a supposed imitation of their institutions, but French liberty must learn to thrive on Chicago pork or be denounced. Germany angers them just at present; but her trade is so poor, and her people emigrating in such crowds to America, that they rather look down on her. Russia may have once been a political ally, but is now forgotten. I doubt if the ordinary man has any greater cognizance of foreign countries. China represents Asia, the negro Africa, the Sandwich Islands represents a sugar monopoly and the world of the Pacific.

Public offices failing to help clear the way, you will fall back on your acquaintances; near the stove in the hall you will valiantly check any nervous apprehension of the habit which demands spittoons, and listen with politeness to the friend who harangues in periods; he is obliging and loquacious, but in the end you will find yourself no further advanced than with the sapient advice, "Go and see for yourself."

For any one who wishes to experiment in the life and habits of the prairies, time spent east of the Missouri is time wasted. Once he has stretched his legs in New York, and got rid of the cramps acquired on board ship, let him take his ticket through to some point, say as far as Denver, and thence make a fresh start. The Eastern cities are not particularly interesting on the face—new, formal, and extensive; to be appreciated, their business aspect must be studied, which requires special aptitude and facilities. Every traveller makes his pilgrimage to Mount Vernon, so called by Washington's elder brother, to whom the estate first belonged, after Admiral Vernon, under whom he had served. Forgetting bygone quarrels, one's inclination is to claim these men of English stock and English names, and to reflect that the settlement of the family quarrel is not one which need now embitter feelings, nor check our admiration of what has been done since a hundred years ago. Washington's house, as well as one or two other country homes we passed, had a strange resemblance to the houses of our people in India—(*Kothi* in the vernacular)—that is, in the appearance of the classic façade, with its straight lines and pillared verandah; the outhouses, servants' buildings, etc., added the similarity of arrangement. Here the material for the walls is wood; there brick; but the coat of white or colorwash covers these and completes the likeness. If I have not jumped to a hasty conclusion from too few cases observed, this resemblance, accompanied by the fact of its departure from any of the ordinary types of English domestic architecture, is remarkable. The climate and circumstances of life to which the Anglo-Saxon in the East and in the West emigrated, in many ways were similar; that is, a hot sun during the summer, in which a European labors at a disadvantage, and a number of servants and workpeople of a humble and easily-managed race, born to toil, and fed at a minimum cost. If, then, every act has a scientific value, have we, in the above case, without going to the lower creation, unearthed a case of adaptability? Is it not also one which might help a naturalist of lively faith to some very large conclusions? Another admirable instance of means suited to an end was the good-sized mallet wielded by the

Chairman of the House of Representatives, with which he successfully drowned all discussion if the gentlemen from the various States were turbulent and noisy. The President of the Deputies in Paris has a bell which is likewise efficacious, and tides him over minor difficulties which do not require the extreme measure of his putting on his hat. The pen is more powerful than the sword, the mallet is more powerful than many tongues, and reduces the House to order when gentler means would fail. A group of boys on the steps of the platform catches one's attention; their duties are fetching, distributing papers, etc.; their occupation wrangling and rough play, in supreme disregard of the dignity of the House, and the prosy arguments of some speaker. They recalled to my memory the antics of a band of choristers in St. Peter's, whose levity and turbulent joy when their services were over, contrasted with their prim demureness while marching through the cathedral, shocked one's sense of awe in the sacredness of God's house, and jarred with the imposing stillness of the temple.

Out of New York the quality of the food fell off badly; even in first-class hotels, the liberal *menus*, with the choice of twenty dishes, where in Europe you find but two or three, so often extolled, are the most laughable take-in; each small article on the table is enumerated, while many of the dishes are repetitions—for instance, under broiled and fried you find the same chops, steaks, and fish spun out into a dozen items. *Relishes* will head various sauces, pickles, and vinegars; eight or ten different shapes of bread are specified; the same for preparations of eggs. Potatoes, boiled meats, cold meats, vegetables, each give an opening for more names; while for drink you can drown care in black tea, green tea, Oolong tea, English breakfast-tea, ice tea, coffee, chocolate, milk. The long list does no harm. The black waiter is an attentive fellow if you tip him well; but the materials, in my estimation, left much to be desired for improvement—at least, I thought so, until my opinions were completely upset by being told that "We Americans never can find anything fit to eat abroad." Then I humbled myself, and recognized the good sense of the proverb "*de gustibus*," for we need only refer to French authors for an opinion on English cookery. The Caribs, we are led to believe, relish cold missionary, and habit may make a people prefer fried meat, hot bread raised with baking powder, and a cup of tea or coffee as the staple of their dinner. This selection, from a health point of view, is funny, particularly as the American is said to have bad teeth and a weak digestion. He eats little; often tearing a few scraps from his steak with a blunt knife, he pecks his fork into one or two of the dozen soap-dishes in which the meal is served; he swallows hastily a cup of strong coffee, and leaves the table. In the hall he will drink a tumbler of ice-water, and then start a smoke or a chew; this process may stop hunger, but is not dining, would surely be a Frenchman's comment. In every art of civilization there is a descending scale; and while east of the Missouri the bottom of culinary talent is still a long way off, a thin, leathery fried steak and a piece of stringy lamb will smile as dainties after a course of doughy rolls and coarse bacon; but to this we have not yet come.

Really, after giving advice to go West at once, my narrative jibs terribly at starting; as it takes time to get over the distance, perhaps a too hurried skip from New York to Colorado would be artistically incorrect; it might leave an impression at variance with the interminable hours of railway travelling, notwithstanding that a good many are spent in sleep, which intervene. A sleeper, or berth in a sleeping-car, is no extravagant luxury for any one who has to travel more than twenty-four hours; the ordinary seat in an ordinary car forbids comfort and defies sleep. By no arrangement of head, body, or limbs; by no propping with luggage, nor spreading of coats and rugs; neither by resting your head on the sharp angle of the window, nor by trying to balance it on the back or arm of the seat, with your legs stretched out under the bench in front, or bent upon your own red-plush mockery of a seat—in no way can you be at ease. The place is too short, and too narrow; for two fair-sized men sitting upright there is no elbow-room to spare. The constant passage of the guard and brakesman down the central aisle, passengers coming and leaving, the call for tickets at each change of train-staff—say every three hours—combine to transform a night-journey into an ingenious system for keeping you awake. Two whole seats, if the passengers are few and the guard sufficiently condescending to care for your comfort, improve matters somewhat; but even then the most experienced drummer, as a bagman is called in the States, makes but an indifferent bed after he has lifted out the cushions and laid them longways, blocked a corner with his hand-bag, and packed away his superfluous length of arms and legs as well as a life half spent on the road from town to town may have taught him.

Therefore, if you have the money, take a sleeper, and if you wish to enjoy your journey travel in the Pullman throughout. As for the theory of equality which used to be illustrated by the doctrine of a one-class carriage for all comers, that has quite changed. There are Pullman cars, first and second classes, cars for negroes or Chinamen, and emigrant trains. Further, you can slip down on a luggage-train by feeing the conductor; or by some process equivalent to riding on the buffers you may do as the boot-black boys, who make their journeys for nothing at all. These two last modes of locomotion, however, are not recognized by the company, and the traveller without a ticket or currency is sometimes shot out of the train after it has been brought to a standstill in the most lonely and desert spot procurable. The journey is, as a rule, monotonous, in spite of the disturbing influence of the fruit-vender, and the best endeavors of your interrogating companions. The latter are often good fellows; and if you have no important business you can forgive readily enough the personal interest of their questions; but the man who sells things is an enemy. First he offers you yesterday's newspapers; next time he walks through the carriage he drops two or three handbooks, guides, maps, or magazines beside each traveller; the next trip he forces the choice of apples or pears, then oranges, California grapes, dried figs, maple sugar, including an advertisement; cigars—each item nearly requires a separate trip up and down the carriage ;—last, he brings his basket of pea-nuts, and throwing two or three into every one's lap, he has completed the round; not for the day, but for the nonce; he will begin again at the beginning, for sure, after dinner. This itinerant trader certainly should be suppressed; his prices are extravagant and his office unnecessary.

In the mean time the train rolls on, at no frightening speed, though advertised as a lightning or thunderbolt express. Twenty miles an hour carried on all day soon mount up into very respectable figures; and travellers who are charmed by a big total of movement delight in the hundreds of miles reeled off and entered in their log. To some people, however, life seems wasted either in a ship or on these long train journeys; for what pleasure may be found in the contracted space, the numerous meals hurriedly gobbled, and uniformly regretted; the constantly passing landscape which leaves no further impression than so many miles of sea; the chatty traveller; the smoking traveller; the man who asks you to cut in at whist; the silent man, and last, greatest bore of all, your ill-at-ease self, whom you can neither forget, employ, nor avoid. At last you run into port or your terminus, one last concern for your luggage, of which in America you are relieved at the fancy price of a shilling an article; a seat in the bus costs one or two shillings, and so rejoicing you return to the ordinary cares of life.

Colorado is a State more known by name to English folk than many older ones; settlers, sportsmen, cattlemen, farmers, miners, invalids, have all had a say on the subject; the climate, the mountains, the scenery and resources have all been praised. Many have found a home, a few fortune, and some have left their all in the Centennial State. There is still room, and whether you wish to amuse yourself, or to find occupation, or to try a new life, there are openings; the only thing is to be careful, and in no hurry; settle and wait. If you wish to emigrate start young. The Americans tell you that all Englishmen inherit money from their parents, for which information they quote their newspapers as sufficient authority. The great purpose of Nature is to relieve this plethora by transferring the cash into the pockets of Jonathan; this is called developing our resources by Eastern capital, and consists mainly of selling land, stocks, shares, etc. Therefore, to be on the safe side, if you have capital, leave it at home; learn the business you wish to follow by working at it with your own hands; pay no premiums, but hire yourself out; if active and willing, you are well worth your keep, and in a couple of months, if a sensible man, and meaning to get on, your employer will be glad to give you wages, for steady men are scarce. Many know their work, few will do it, still fewer are to be trusted out of sight. You will soon be able to save money—a very little no doubt, but enough to make you think of investing. This will set you inquiring into prices, and the chances of a return; you will probably by one or two bad deals pick up experience, and learn that saddest lesson—a distrust of men. After a couple of years you may venture an independent start. You can, by accepting American citizenship, take up 160 and 320 acres at very little expense under the Government land laws ; or, if disinclined to change your state, you may buy out some one else who has pre-empted a claim suitable to your purposes ; your money will help you to stock it and buy farm implements. The tenderfoot who takes his dollars in his trouser-pockets is a lost man. Every old settler with a poor farm, a worn-out wagon and horses, a valueless mine or property, will make a dead set at the coin, and they are not easily to be shaken off. A gentlemanly looking man will, five minutes after a first introduction, offer you shares or an allotment, or invite you to join his company, or to try a plot in his new city ; a refusal does not end the matter if he knows the money is there. The hotel hall is a public place, and one in which business is usually carried on ; after dinner he will sit beside you, and suggest your going down to see the property ; next day he will introduce a gentleman who knows all about it ; you have now two to deal with, for they are the most neighborly people in assisting a friend to clinch a deal. If you can hold your own, congratulate yourself—you are more fortunate than most.

In Denver you are in the middle of the stock country: north, south, east and west, cattle have been raised and are still running on the prairies where the grass has not been fed off. The prairies include all the unsettled parts ; they are sometimes grass land, sometimes covered with sage and other brush, amongst which grass is found ; the term takes in flat table-lands, the slopes of mountains, and what are called bad lands, which are the wildest jumble of hills, ravines, small flats of excellent grass, and stretches of almost bare lava rocks. The name is derived from the French, who wrote on their maps *terres mauvaises à traverser*. In the hurry of business the two first words only were translated, and consequently left a wrong impression, for these lands often afford excellent cattle-ranges; the grass is rich, water is to be found in many deep ravines, the broken ground gives good shelter against storms, and the farmer will never come to oust you. The men working on the prairies are to a large extent young Americans who have left Missouri, Illinois, Iowa, and other Western States to seek their fortunes, with a certain sprinkling of foreigners; but the numbers of these are not so marked as in the longer-settled States, where farming and commerce attract emigrants accustomed to those industries, or even as in the mining camps. The usual ranches are poor buildings, built like the Russians do their houses, of logs joggled at the ends, but here they are very inferior, and seldom have more than two rooms; even one room is common; the ready excuse for not improving his home is that it is temporary, though the man and his family may have occupied it for several years. The stable is a mere shed, the walls of thin poles badly put together; it is hardly ever cleaned out, the animals stand in pits stamped in the heap of manure; but the horses generally run loose, and it is only while the team is working that the stable is occupied. At the cattle-ranches, where half-a-dozen cow-boys may have to spend the winter, and for which more money is forthcoming, fair-sized rooms are put up, the accommodation being increased and improved year by year. Bunks occupy one end of the room, a huge fireplace the other, from which the mound of hot ashes, topped by two enormous logs, fills the room with light and warmth. A large area of ground is fenced near the ranch, in which horses likely to be required are turned loose, or portions reserved for cutting hay. The range lies outside this, its extent depending on the cattle man's ideas, tempered by the opinions of his near neighbors. There are of course no absolute rights; the land is all Government, even probably that which is fenced, and there is little attempt to segregate the herds. Some of the territories have passed laws acknowledging settler's rights on streams or to pieces of land they have enclosed; but this is contrary to State law, and latterly a circular was issued in California pointing this out, and distinctly laying down the law that others could enter on such land without trespass. Among stock-raisers, however, there is much give and take: the first settlers naturally try to keep out newcomers; they must end in accepting the inevitable, and that is, so long as there is grass cattle will crowd in. But the greatest enemy to stock is the plough: the farmers are coming slowly but surely from the eastward; parts of Kansas and Nebraska have gone over to tillage; stock must give way and disappear into the mountains and rugged country. Almost the whole of Wyoming, Montana, and Idaho are still unsettled, and in these territories the cattle business is still carried on somewhat in the old style. Formerly the man who shouted loudest, galloped hardest, and was quickest in drawing

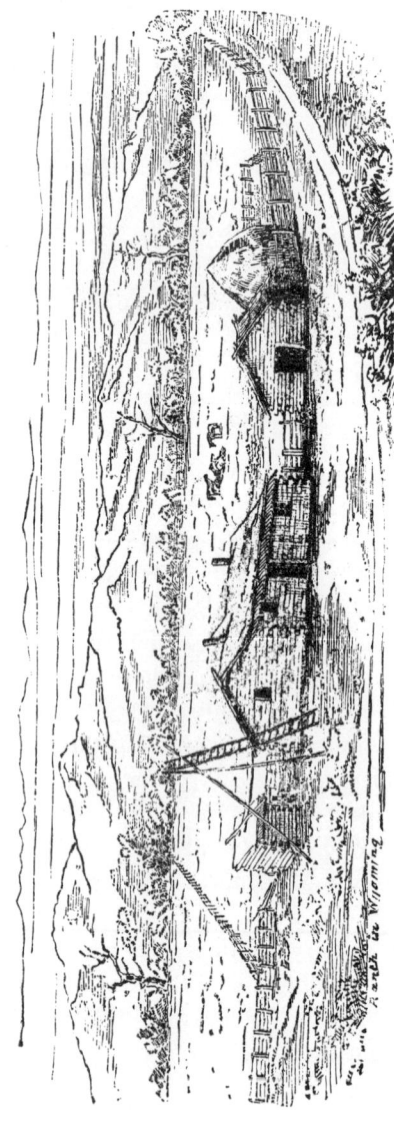

RANCH IN WYOMING.

his "gun," was considered the most dashing cow-boy; if he had come up on the Texas trail, and had failed to kill his man, he was held to have wasted his opportunities. But times are changing; it is only in the south, for instance Arizona, where the term cow-boy is equivalent to desperado; in the north the men on the ranges are as good as any class of Americans. The increased value of the cattle has introduced more care and gentler handling in their management.

While roaming on the range the less the cattle are interfered with the better, particularly in the winter. In this half wild state they can take much better care of themselves, and find shelter and food, whereas, if they were herded, that is, controlled in any way by men, they would probably starve. The cows, which in cow-boy language include all sexes and sizes, split up into bunches and take possession of some small valley or slope where water is procurable at no great distance; the shallowest spring bubbling up through mud will satisfy a small lot, if they get it to themselves—a spring so small that, knowing it must exist from the presence of the cattle, you would scarcely find unless for their tracks, and when you reach it there is nothing fit for you to drink, and likely your horse will refuse the mixture of mud in alkaline water which pleases the cow. If water is scarce the cattle must make long tramps, and the country is then crossed by deeply trodden paths, which are an unerring guide to the thirsty horseman; the cattle come down these paths just before the sun gets hot, have a drink, and then lie down till the evening, when they go off again to the pasture at some distance, and probably feed most part of the night. In the beginning of winter the cattle leave the high ground, and the appearance of a few hundred head in the valley which the day before was empty tell the tale of severe cold or snow-storms in the mountains; they like the shelter of heavy timber, which is found along the banks of streams, and here at some rapid or at the tail of a beaver dam is their latest chance for getting water. They cannot, like horses, eat snow, nor does their instinct suggest to them to paw away that covering to reach the grass beneath; in fact, the cattle will sometimes attach themselves to a herd of horses, sustaining life by following in their footsteps. When times are hard the cattle will subsist on grease-wood, and eat almost anything, but till the young sprouts begin to shoot there is on the prairie little to find after the snow covers the ground; bare cotton-wood trees line the streams, on whose bark horses will manage to keep alive, but the cattle are far less hardy than horses; these will come through the exceptional winters in tolerable condition, when twenty per cent of the cattle have been lost.

But in choosing between breeding cattle and horses, the former have some advantages. There is, and must always be, an increasing demand for beef, and in the disposal of your live stock there is a great convenience in being able to ship any number by a train load to Chicago, and there disposing of them in a day; whereas with horses they must generally be got rid of in small lots; there is besides more trouble with horse-thieves, both red and white, than with cattle-thieves. The

Indians often shoot a calf for food in the winter, being altogether easier to find and better to eat than deer or buffalo; some stockmen believe they lose a large number from the redskins, but these are decreasing in numbers every year and are continually being restricted to smaller reservations, in which they are more successfully watched, and by which they are degraded to beggary. It was rather humorous to read occasionally a paragraph in the newspaper, telling of a "new treaty between the Government of Washington and the chiefs of a certain tribe, in which the latter had been induced to give up two thirds of their location, thus restoring a large section of valuable land to the enterprise of our citizens, and still maintaining the best interests of the tribe itself." The Indian policy of the United States is philanthropical in theory; the theory belongs to the East, but in the West its practical settlement is carried on. The white man says, "I have no use for the Indians," meaning he wants nothing to do with them— but he covets their land, and is continually agitating to have the Indians moved further off and the reservation thrown open to settlement; he objects to the Indians being allowed to hunt outside their limits, on the ground that the game in the latter place has been preserved in their interests, and the white man is on that gronnd treated as a poacher. Though loving idleness himself, the settler hates it in the Indian; perhaps it is envy that the latter seems better able to carry out the system of living without working, though any one passing one of these noblemen of nature on a cold day, with the wind cutting the skin, might well pity the poor fellow, whose gaudily-colored but loosely-woven blanket and deficient clothing protect him nohow, while the skirts are fluttering and exposing odd corners of his body to the keen blast. One man suggested compulsory European garb as a remedy, for "the Indian could never work so long as it required two hands to hold together his clothes." There was sense in this remark. However we may feel on the subject before entering on a Western life, one soon joins the opinion of the majority that the Indian is in the way; he is, however, doing his best and removing himself as quickly as the least considerate could fairly ask of him.

To obtain any knowledge of life on the prairie you must cast yourself loose from the railway, choosing for preference some fair-sized town at which to make a start, as a considerable amount of preparation is necessary in the way of carriage, utensils, and food. If you are in a small place with a single store you will very likely have to pay 20 per cent more all round, and be delayed as well. If going north of the Union Pacific, Cheyenne might serve better than most places in which to make preparations. If you have never travelled before on the prairie you will be much puzzled to know what to take and what to do without. As to the former, a beginner must be guided by some experienced person, who may be hired to accompany him; as for personal effects, these will have to be curtailed considerably; anything in the shape of a box should be rejected, for even should you start with a wagon, your luck, or the want of roads in a part of the country you would wish to penetrate, may force you to pack your kit on horseback. In such matters detailed advice is valueless; each man has his own idea of comfort and of what is indispensable. It mostly happens to every man by degrees to throw away a good deal of his kit after he has been out a month, or to deposit it in some small place of safety, where it will possibly remain for all time, so far as he has a chance of recovering it. It is not given to everybody to be really independent. The Indian mounts his pony, and, burdened with nothing but his gun, will travel any distance, finding a meal here or there in some camp, or going without with lofty disregard of his stomach. The experienced frontier-man can very nearly rival this: when travelling a trail he knows something about, a good horse is enough; a couple of blankets, one under the saddle and one strapped behind, is all his bedding; his clothes are on his back, his food he will find in some cabin or hunter's camp of which he has been told the situation. Avoidable hunger is to most men's notions a deadly sin. The freest way to travel is to have two horses, one to ride and one to pack; with these you care not for roads nor habitations; you can carry enough food for a fortnight, and travel thirty miles a day; water and grass for the horses are your main care, and next to those fuel. Not that thirty miles need tie you, for journeys of double that distance or more are of daily occurrence; it is better to make a push and reach good quarters than stop on the way and starve your beasts. It will, however, be some time before your training will permit of your going across country quite alone. If your means do not allow of your hiring a man, you must look out for some other outfit taking the line of country in which you are going and make some arrangement for travelling together.

The plan of the party I joined was to leave the railway at about Rawlins, travel northwards till we reached the Big Horn range, and thence take a trail northwestward into the Yellowstone country : this would traverse a great deal of wild country where there are no settlers and, excepting a few cattle-ranches, no houses at all ; the trail is passable by wheel traffic, so our luggage and provisions were carried in a wagon. The luxury of a wagon will be soon found out once you have tried packing ; you tumble the things into the wagon in a few minutes, while packing the most moderate kit takes time ; you take nothing out of the wagon-box but what you require, whereas at each halt everything must come off the animals' backs ; if, therefore, there is any chance of a wagon being able to get through, there should be no hesitation in preferring it. Once the weather is settled life on the prairies is very pleasant—the work is not hard, the meals are sufficient, and the exposure all day long hardens the constitution and enables you to sleep in the open and disregard an amount of cold which is at first unbearable to the skin made sensitive by indoor life and warmed rooms. The first break of day wakens the camp. You tumble out of bed into the crisp morning air and make a fire ; in about an hour's time breakfast should be over ; the horses are then caught and harnessed, the cook has washed up and packed the wagon, he climbs on to the box and the others mount their horses ; you are off. There is not much to do while keeping along the road ; the game has been probably driven off, but it is a pleasant life ; existence is happiness when the body is in good trim, the mind free from care, and the sun shining brightly : the clear air is exhilarating and sharpens the senses. You feel your day well spent and your conscience approving when your horses have covered a good distance and you have satisfied a keen appetite for the third time; you can take a last look at the animals to notice in which direction they are feeding, and then close up to the camp-fire till it is time to go to bed.

So long as we keep the road the wagon gets along all right, but after a few days we must diverge on to a branch road which crosses the Sweetwater; it is not now such plain sailing, and constantly the wagon must be led around to avoid the washouts across the trail. To drive across country a wagon fairly loaded, with four good horses or mules, is no light job; there are plenty of ugly places which must be tackled; it requires on the driver's part nerve to face the difficulties, experience in passing through them, and caution in avoiding a smash or upset, for the whole outfit depends on the wagon, and though minor damages may be repaired or some makeshift adopted, a serious accident cripples the expedition. The first day's drive after a long rest is always adventurous; the team are fresh, and apt to meet a check in the draught by a dash forward; four good animals pulling with a will have a wonderful way of starting something, whether iron or wood, should the wagon be held by a dip or rut. It is often indispensable that one rider pilot in front, choosing the best places for the wagon to cross any obstacle; even with this assistance the box-seat will test a man's skill, and no one need venture to occupy the position unless he has something more than good-will to recommend him. Should the wheels on one side only drop into a good mud-hole this will afford ample occupation for one or more hours; it is often the best plan to unhitch at once, take the team round to the hind end of the wagon, and hook on to the hind axle; one man must then hold the pole to steer the wagon; the horses may with a good pull clear you; if not you must unload, and even sometimes dig. This occupation is not to any one's liking, and if the accident appears unnecessary you are much inclined to grumble at the driver, who will naturally abuse the man who should have piloted him; you make a short journey that day, camp somewhere uncomfortable for want of water, fuel, or feed, and are generally reminded of the fact you were forgetting, that there is no Elysium here below.

The trail lay through the Rattlesnake range, which was crossed by descending a striking-looking canyon; water was always scarce, and the springs we found were strongly impregnated with alkali; the water is disagreeable in flavor but seldom troublesome to digestion. There are plenty of antelope about, but no other game, unless it be sage hens. Our road takes us across Poison creek and Bad-water, after which it rises steadily towards the Big Horn range. There has been a great lack of invention in naming the streams throughout these western territories. Some few names are repeated again and again, attributed either by a quality, e.g. Sweet-water, Bad-water, No-water, or by a natural feature of the banks, such as Cottonwood creek, Willow creek, Grass creek; or the name may derive from some common animal, as Bear creek, Antelope creek, Beaver creek. Such names recur so often that they lead to confusion, and it is quite an advantage to find a distinctive title, if not more resounding than Tin-pot or Deadman, or one taken from the patronymic of its discoverer, or of the first settler on its banks. We were now in the middle of the cattle country, and having encamped, there was an excellent opportunity of seeing how the round-ups were carried on and the cattle worked.

A round-up is the general arrangement among cattle men in a given district to work the cattle by a common establishment: each owner sends one or more cow-boys to represent his brand and to take charge of all animals belonging to his herd. The management is placed in the hands of some experienced foreman, and the ground to be covered is of great extent, occupying the men from a couple of months to a season. The main plan is each day to drive the cattle out of all outlying valleys into some central level spot; out of the mixed mass the different brands are separated, beginning with the largest herds. This is a distinct advantage to the large owners, as the principal object of the general round-up is to get at the young calves. While these are being cut out, as it is called, the cattle in the main bunch are churned up, so that calves get separated from their mothers; and as the only title to a calf is that it is following a cow with your brand, those who cut-out last will naturally lose some which belong to them. Any unbranded calves which are not following a cow are called "movricks," and belong either to the man on whose range they have been found, or are shared, according to the local custom. The process of cutting-out a cow and calf is very pretty if neatly done; one man can do it; with two it goes easier. The cow-boy rides through the gathering of cattle until he sees a cow and a calf belonging to him. He follows these quietly, trying to shove them towards the edge of the herd. As he gets them moving he quickens his pace, and when on the outside he will try to push them straight out of the mass; but the cow is disinclined to leave her companions, and generally tries by running round to break back into the main bunch. This the cow-boy has to prevent, by riding between the cow and her object. Cow, calf, and horse are soon going their best, and the cow-boy must be ready to turn as quickly as his game; he must, however, be careful not to separate the young one, for should this happen his labor is lost; he must let the cow rejoin the herd and recover her calf. Each batch of cows thus separated is kept at a certain distance off, say 200 yards, and is watched by a man to prevent them rejoining the main herd, or from mixing together. If the cow-boy has been successful, the cow is soon blown, and, finding herself checked in doing what she wishes, will yield; and seeing another lot of cattle which she is not interrupted from joining, will trot contentedly towards them, and having her calf alongside will settle down quietly.

This cutting-out goes on all the day long, until the whole herd is divided. It is hard work on the men, and particularly so on the horses, which have to be changed two or three times during the day. The quick turning and stopping must shake their legs, and certainly brings on sore backs. Their mouths do not suffer; riding with very severe bits the cow-boy has necessarily a very light hand, and hardly uses the reins for turning; the horses know the work, and a touch on the neck brings them round at a pace which sends the beginner out of his saddle. The cow-ponies are rather small animals, and half disappear under the big saddles of the cow-boys, which often weigh forty pounds. The origin of the cow-ponies is the bronco, which came into the country with the cattle driven up from Texas; they have, however, been much improved in latter years. The biggest are by no means the best; a short, compact pony of about fourteen hands works more quickly than a larger animal. Some of them, with small, well-shaped heads and bright eyes, are very taking-looking animals; their manes and coats are shaggy, showing coarse breeding, and their tempers not to be trusted. Each boy, when out cow-punching, rides from six to ten horses, using them in turns, and without the slightest compunction riding one horse fifty or sixty miles, of which a good deal may be fast work. After the day's duty he takes off the saddle and bridle, and without further ado lets the horse loose, who, after a good roll, takes up the scent and rejoins the herd of horses; his turn for work will not come round again for several days. Of course they get nothing to eat but the grass they pick up; they are seldom shod. Their half-wild origin is attested by the majority of duns and sorrels. The heavy saddles are believed to be on the whole an advantage, as from their size and solidity they distribute the weight of the rider and his kit over a larger portion of the horse's back. There is truth in this; and for long journeys probably the ease of the big saddle more than compensates for the extra weight; but in roping cattle the heavy saddle is absolutely necessary. There are often two girths; these must be well tightened, and even then the jerks try the horses severely. The end of the rope is held fast by a turn round the horn, which stands six inches above the pommel; the rider has often to hang heavily over the further side to prevent the chance of the whole saddle being turned round. The big spurs do not hurt the horses; to make them effective at all, the cow-boy reaches his heels forward, and spurs his horse in the shoulder.

If there is still time, it is best to brand the calves the same day, as, after that operation, the cattle may often be turned loose to run on the same range in which they were caught; but if the outfit to which they belong has its principal range at some distance, the batch must be taken off, and driven and watched till they arrive on their own range. It is not absolutely necessary to have a corral to brand in, but if you can run your bunch into one, it saves trouble. The corral is roughly and strongly made of posts and rails about five feet high; it should be big enough to hold your bunch of cattle, and leave room for working. Just outside a fire is lit, and one man keeps the brands hot, which he passes through the rails as they are called for. In a small corral one man on horseback is enough inside, and he can be dispensed with unless there are any large calves to handle. A man, armed with a rope-lasso, catches a calf by throwing it over his head; if a little fellow, the calf is dragged to one side, caught and thrown down, cut and branded in a very short time; but a calf of two or three months even is not so easily managed. The noose having been tightened on his neck, the end of the rope is passed round one of the rails; the calf gallops up and down the arena at the fullest length of his tether, jumping and bellowing as if he knew his end was coming. By degrees the rope is overhauled, and the length which gives the calf play is shortened. One of the men will then go up to it, catching it by the rope round the neck in one hand, and, passing his hand over its back, by the loose skin on its flank near the stifle, with the other. The more the calf jumps the better, and if he is slow and stupid he will get a shake to rouse him. Taking the time therefore by the calf, the man seizes the opportunity of one of his prances, puts a knee under him to turn his body over, and then lets him drop to the ground on his side. Another catches hold of a hind-leg, which is stretched out to its full length; the first sits near its head, with one knee on the neck, and doubles up one fore-foot. The calves generally lie quietly, and do not bellow even when they feel the hot iron; but a few make up for the silent ones by roaring their best.

A good-sized calf gives a lot of trouble. After the rope round the neck has been drawn up, another noose is thrown to catch one of the hind-legs, which should be the one not on the side to be branded. This rope is also passed round a rail, and hauled tight till the animal is well extended. Somebody takes hold of his tail, and

LASSOING ON THE PRAIRIE.

with a strong jerk throws him on to his side. A hitch is taken with the same rope round his other hind-foot; the noose is loosened round his throat; but the man leans his best on his neck, and holds his foreleg tightly. He must look out for the brute's head, as the calf throws it about, and if it should strike the man's thigh instead of the ground, as it is very liable to do, he will receive a bruise from the young horns which he will not have the chance of forgetting for a good many days. The brand should not be red-hot, and when applied to the hide should be pressed only just sufficiently to keep it in one place; the brand if properly done shows by a pink color that it has bitten into the skin, well through the hair. Some of the stock, in the early spring, have very shaggy coats; and a brand applied only so long to their hide as would answer in most cases, would leave a bad mark which would hardly show next winter. The calf when finished with generally gets up quietly, so soon as it feels the ropes loose, and rejoins the others. The cows seldom interfere to protect their progeny; when you do find one on the war-path, it makes the ring lively, and all hands are prepared, at short notice, to nimbly climb the fence or jump over.

To keep steadily at catching, throwing, and branding is hard work. The sun is hot, the corral full of dust from the cattle running round and round, and your clean suit is spoiled with the blood and dirt of the operations; you may have besides a tumble yourself when throwing a calf. The process is still worse if rain has fallen, and the cattle have probably for want of time the day they were corralled, been kept shut up through the night. As they run round and round to avoid the man they see swinging his lasso, the whole area is churned into mud; the animals dragged up get covered with filth, which is passed on to the men at work. There is a certain excitement about the business; the cow-boys will work at it very hard and through very long hours. The boss is a great sight, and never tires, running backwards and forwards between the fire and the struggling calves; each time he slaps on the brand he seals a bit of property worth ten to fifteen dollars—he would like to work at this all day long. If the corral is very large the ropes are thrown by a man on horseback. So soon as a calf is caught he takes a turn with the end of the rope round the horn of his saddle, and the horse drags the animal to the right spot. A cow accustomed to men on horseback will sometimes run after her calf with her nose stretched down towards it, no doubt inquiring the nature of its trouble, and a "What can I do for you?" but so soon as she nears the men on foot the cow stops, and then leaves the calf to its fate. If branding is done in the open, one man holds the bunch together, and the lassoer picks out the unbranded calves, and drags them off to the fire. If large cattle have to be branded you can do nothing without horses. The lasso should be thrown over the horns only; it takes three or four men to hold the animal after it is down. When it comes to an old bull, and he declines to be maltreated, he has his own way. A couple of ropes thrown over his horns and tied to a post he snaps like a packthread. A brand can be put on him by a man on horseback, with a hot iron in his hand, following the bull into the thick of the herd; jammed in a corner of the corral the bull can move but slowly, and there is time to press the brand, and to leave a mark. Throwing the big cattle does them no good. For all purposes it would be a better plan to arrange the corrals with pens and shoots for both separating the different brands, and for doing the necessary ear-cutting, branding, etc.

When the cattle in one place have been settled with the round-up moves on; the camp is broken up, wagons packed, and a string of four-horse teams make a start. The cow-boys, with their *schaps*, i.e. leather leggings and flopping wide-brimmed hats, are trooping off in different directions, puffing their cigarettes, and discussing the merits of their mounts. On both sides moving clouds of dust half-conceal a mob of trotting horses, which are the spare animals being taken along to the next halting-ground. Soon the place which was lively with bustle is left desert, marked only by the grass trampled down and the heaps of dirt round the old camp. The cayote will sneak in, and have his pickings on the offal, scraps of leather or ends of lariat; then all will be quiet till the autumn round-up, or even till next spring.

After the calves the fat cattle have to be separated from the herd and driven off in the direction of the railway; this drive may occupy one or two months, and must be done with deliberation and quietness. The seed-bearing grasses are very fattening, and the tendency of all the cattle is to grow rounder and more sleek till late in the autumn; this condition is natural and very necessary to enable them to live through the winter. The steers, mostly three and four years old, having been collected into a band, are moved slowly from day to day, care being taken that they cross plenty of grass and water. At first they are wild, and even the men on horseback have to hold back a little distance, showing themselves just enough to keep the herd headed in the right direction. All galloping or shouting is discouraged; nothing must be allowed to startle the steers; a man on foot would possibly drive the whole herd off into a mad stampede. A few old bulls past work are often included in the bunch of fat cattle. A low price is paid per pound for them in Chicago, but they weigh heavily. It is said that they are made up into the preserved meat in tins. On the trail they are useful as setting an example of steadiness. If the steers are kindly handled and not over-driven, being young, fat, and frisky, they are ready to romp; should they stampede, the bulls, heavy and old and not easily scared, hang back, and look about for the cause of the run. They will stop, and the steers near them will follow suit. The cattle of the northern territories have the character of being easily stampeded, but they seldom run far; on the other hand, the Texan cattle go for miles.

One or two men must be continually in advance to drive off the range cattle, who might otherwise mix themselves with the steers, and give much trouble in cutting them out. On the actual journey the herd is encouraged to string-out; the leaders find their place every day, and it is only necessary to keep them along the right trail. A boy on either side, and two at the end to work up stragglers, are sufficient, though the line may be over half a mile long. When halted to feed, the herd should be surrounded, half the men doing this work in turns, the other half getting dinner, if in luck's way; but as it is necessary, both for food and to avoid disturbance, to take the cattle by the most unfrequented routes, the wagon may have ten miles to go round in addition to the march of the herd. In these cases breakfast must last till supper-time, except for a snack that the boys carry with them. It is important that the herd should never be left unwatched. When at night it is thought time, they are driven on to a bedding-ground and bunched up. So soon as they have steadied down, one or two men are left on watch, whose duty is to ride round and round the herd, and prevent any straying. If the weather is not too cold, the night-watch not too long and the cattle behave well, this is not disagreeable work. The cool air is refreshing after the long day's heat and glare; you walk your horse at a little distance from the cows, with an occasional short scamper after some rebels; you must, however, keep moving, and show yourself constantly on all sides. To hear the human voice seems to quiet the cattle, and the man on watch will often sing or call quietly. One by one the animals lie down. You hear a great puff as if all the wind was let out of a big air-cushion; it is a steer settling down on to his side; more puffs, the shadows sink low, and at last there are none left standing. The quiet of all these huge animals is impressive, and seems in keeping with the sleeping earth and calm sky; the voices of the men in camp hardly reach you; a flicker from the fire càtches the higher part of the wagon, and just marks its position. Provided nothing extraneous disturbs the peace, the cattle will lie still up to eleven or twelve o'clock of night, while you circle in the darkness round the black patch on the ground, keeping a sharp lookout for any shadowy objects sneaking off in the gloom, and often riding to investigate a suspicious object, which turns out to be only a bush. Before midnight, under some special ordinance of nature, the cows are restless and get on their feet; a few will try to feed out; these you must drive back again. But before that time, if holding the first watch, you have probably been relieved, and are back in your bed. Each man has a horse saddled and picketed near the camp all night; as if anything frightened the herd, or a storm came on, all hands must turn out and mount. If the cattle are really away, you must be after them without delay, and, so soon as you can stop them, bring them back to camp, provided always you know where it is. Any one left behind will make a good bonfire to direct the boys; but a dark night with rain prevents your seeing far, and the camp has often been chosen in a sheltered spot, which makes it more difficult to discern the blaze. The main thing is to keep the herd together, whether still running or halted. If matters have been well managed, and no serious disturbance has occurred, the herd wakes up and starts out at daylight. You string them out along the trail, and take a count, or look that all the bulls and other animals with distinguishing marks show up present to the roll-call, and move off on another day's expedition.

When approaching the railway station at which the

steers are to be shipped, three or four days' notice will secure you a train. At the appointed time the herd is driven into the railway stock-yard. This is a large inclosure, with passages communicating with pens which hold just the number you can cram into a car; the pens are placed at exactly the distance apart of the length of a car. When the business of loading is commenced, the pens are filled; the steers are driven up a shoot and enter the cars; the last one or two have to be prodded and forced to find themselves room. They should then all be fairly distributed throughout, their heads up, and legs clear of each other. A cow hanging its head will get its horns entangled in some other's hind-leg, and when the head is lifted the leg must come too. A steer may often be seen caught by a hind-foot over a rail five feet above the floor. This has happened in trying to kick itself free from the horns of a brother in difficulty; and until the foot was pushed out there it must have remained. When all are properly disposed the bar is dropped, the door shut, and the next pen is emptied into its car. The top of the palisade of the stock-yard is planked, so that you can walk all round and look down on to the cattle. So soon as the last ones have been cooped in, the bell rings, and the train starts. On well-arranged lines the cattle-trains are run as fast as any, and are allowed to take precedence of most other traffic; but every day the train must halt, and the cattle be taken out for several hours to feed and water. At most large stations there are cattle-pens with water running through them, and deep mangers filled with hay; the cattle get a chance of eating and quenching their thirst.

On first getting out of their cars they are more inclined to lie down than to do anything else, for while travelling they are so crowded that they get little rest. As for lying down in the car, that would never do; and during any halt of the train the boys accompanying the herd must take a look round, and, with their poles, prod any cow that is resting, and force it to get up. This is done in their best interests; for any animal once down cannot rise, and is almost sure to be trampled to death, missing the ultimate glory of becoming beef; the carcase is thrown out at a siding and eaten by hogs. The work of loading and unloading along the journey is very expeditious. The new experience of being cooped up and shaken, or some instinct of their impending fate, has sobered the steers; they are no longer the sleek, shining, frisky inhabitants of the prairie. Bones begin to show; their hides are dirty from close quarters and lying down in pens; they cannot eat food enough in the short time at their disposal; their sides flatten, and they walk in and out of their cars with the utmost docility. Twenty minutes are enough to load up a train of two or three hundred beasts; each day the proper number fills into the car with less squeezing. Any delays are consequently annoying to the owner, who hates to see his cattle shrinking. Every pound of flesh lost is money out of pocket; but so long as Chicago is the main market for cattle, they must travel six or seven days by rail from the railway point nearest to their range. There is a great opening for improvement in the meat trade; at present the cattle are sent on the hoof to Chicago and other towns, losing flesh and being bruised on the journey, besides travelling in the most expensive style in which goods can be moved. All this would be obviated if the steers were butchered on the prairies near some railway-station; the steers would be in their primest condition, and only the paying portion of the whole weight would be railed. The cost of refrigerating-cars to carry this meat must be less than the cattle-pens which take a corresponding number of live animals. The commission and expenses along the line and at Chicago would be saved. The meat trade is, however, in few hands and tends to be virtually a monopoly. It would require a very strong company to ship meat directly from the prairies to Liverpool. They would find considerable opposition from all vested interests, but should expect some support from the cattle-owners; though probably the whole success of a new experiment would hinge on the railway, who would certainly differentiate rates and charge a little more for carrying half a bullock weight dead than for the whole when living.

The railway journey is as uncomfortable as it possibly can be to the men accompanying the herd. The only accommodation is the guards' van, which is often crowded by railway work-people and travellers by favor of the conductor. The servants of the railway are often disobliging; and the mere fact of the cow-boys being necessarily of secondary consideration to their charges makes the trip a disagreeable one. The night is no time for sleep. At each halt you must jump out, one man with a lantern, both with goads, walk along the rough ballast, and peer into each car to discover a cow which requires stirring up. Having found an offender, you poke her, prize her, twist her tail, and do your utmost to make her rise. In the middle of your efforts the bell rings, the train starts; you clamber up the side of the wagon on to the roof, and when there make the best of your way back along the top of the train to the rear van. This little trip in the dark is not one to enjoy. There may be twenty cars, say forty feet long each. Before you have crossed two or three the train is going at full speed. Only one man has a lantern; you are incommoded by a heavy great-coat, as the air at night is keen; the step from wagon to wagon requires no more than a slight spring; but it is dark, or, probably worse, the one lantern is bothering your eyes. The rush through the air makes you unsteady; no doubt your nerves are making your knees feel weak. It is a hard alternative to get back to the guard's caboose, or to sit down in the cold on the top of the train until you reach a halting-place; having tried both, it seems that neither can be cheerfully recommended. If you do not climb on to the roof you must take your chance of jumping on to the step of the last car as it goes by; this would be the reasonable way if you were allowed to do it, but as the driver does not care to look back, you must consider whether you are sufficiently an acrobat to rejoin.

Having reached Chicago there is an end of the business; the cattle are turned into the big stock-yard, and sold by commission. To visit these stock-yards and the processes of slaughtering are part of the sightseer's orthodox duties in Chicago and need not to be mentioned here. The above tells roughly the general plan of the work with cattle as practised in Wyoming; but to go through the whole process practically occupies the men from April to October, after which time those who are kept on may prepare to settle into winter quarters. If I now skip lightly through the incidents of a summer spent on the ranges, the petty details of our life and occupations may serve to color a picture which is at present barely sketched. After settling down into a standing camp, our time was occupied in bringing the cattle together from all the outlying valleys; occasionally a bunch was driven in which had been collected as the round-up passed away further; every second or third day some calves had to be branded. A morning spent in a long ride through the mountains was always enjoyable. There was continually something to make a change; either the discovery of a bunch of cattle which had succeeded in keeping out of sight heretofore, or your party might alight on a well-grown yearling with unmarked ears and smooth sides. He is a prize; but to lasso and brand him gives the opening for some fun, for he runs like a deer, is as wild as a hawk, and as strong as a horse. There are few good lassoers in this part of the country, so instead of being roped in the first throw, the yearling probably gets away; over rough ground he can travel pretty fast, although the horses can catch him easily on the level. There are many attempts made before the rope is satisfactorily round his neck. He then begins to plunge and jump, while the man who has hold of him keeps the rope taut; the horses understand this, and a well-trained horse will do it of himself with no one on his back. One man should be able to catch, throw, and brand a cow on the plain; but even with two or three men the object is not always accomplished so very speedily. Should one man dismount, the enraged cow makes for him. If the rope is held tight there is no danger outside the ring; but sometimes the rope breaks, or in the charging and shifting the man on foot may get between the animal and the horse; the cow will make a rush, and the man is lucky if he can escape a tumble and a kick. If no other excitement was on hand we had always the satisfaction of slaying rattlesnakes; they were so numerous on the dry hillsides in the sage-bush, that after the first week one hardly took notice of them, unless of unusual size. Your horse would jump to one side, and at the same time you heard a noise as if a handful of coppers were shaken together; this was a snake giving warning. If you got off you would probably find it refuged in a bush. It was very easy to kill them with stones or a stick; the fact of their being twisted up in a sage-bush prevented them from striking. As a rule their object is to escape, and I only saw one snake fight. He was larger than the run of them, and was lying on an open spot; seeing himself surrounded he was very vicious and met the showers of stones by most determined strikes; these strikes do not reach far—a few feet. The poison of the rattlesnake is not deadly; I could not hear from any one of a case in which a man or larger animal had died from a bite. I saw a single instance of a snake-bite, and that was a horse struck in the fetlock. The swelling, which began in the leg, extended up into the shoulder and chest and lower part of the neck, on one side. The horse was left behind, and for over a week was very sick; he could not follow the herd, nor had he strength to come down to water, but after that time he recovered, and a month later I saw him ridden. It is said, however, that the animal bitten never completely recovers its former vigor. The dangers of a rattlesnake bite must be small, as no remedies are carried about to meet such an accident; the medicine reccommended being the universal one of whiskey inside, tobacco outside. There is not much game. We see antelope, who are simple-looking animals till disturbed, when they erect and spread out a brilliant, white, fanlike tail, behind which the animal disappears, all but horns and legs; he goes off with great bounds. We get a few; the meat is very good. There are plenty of jack-rabbits as big as English hares, and, to my mind, better in the pot. Sage-hens might have been easily shot, but their flesh is said to be tough and ill-flavored; they were not at all wild, and ran through the bushes with their brood just out of reach. The old hen is a very courageous bird, and defends itself against other animals rather than take flight. Perhaps due to the immunity from appearing at table, and therefore not often troubled, it apparently does not fear man. One of the party tried to knock over a chicken out of a covey which were running through the bush. After several shots he winged one of the young ones, which began to scream; the old bird rose in the air and boldly attacked the hunter, who could only keep her off by shouting and waving his revolver in front of his face. Higher up in the mountains you might find prairie-chickens and deer, but to reach them would be more than a day's trip from camp. One old buffalo is ridden and killed; his meat, even as an alternative to bacon, does not go down.

The man brought up in civilized ways at first finds himself puzzled in these uninhabited prairies. You start out in company with some one else, and take no notice of the direction travelled; besides, in going outward and coming homewards, objects of course will look quite differently. After getting out some miles, you may separate and each take a round, one bearing right, the other left. You are riding across country, and have often to diverge to avoid unnecessarily climbing a hill, or to seek a ford; if quite unaccustomed to prairie ways, you soon are troubled by a nervous feeling of being turned round, and when new to that bit of country, not knowing the principal features, you are easily in doubt as to your road. You must not get lost, for it depends on yourself alone to find your way back, where your little adventure, if guessed, will be greeted by a hearty laugh at the tenderfoot. The Western man, like the Red Indian or the wild animals, somehow always knows where he is, and has a map of the country in his head. There are many stories of guides, or other frontier men, starting off across two or three hundred miles of unexplored country, to reach some other trail which lies—out there; and this with probably only the most slender outfit to just last the time, on the narrowest scale, if all goes well. Add the chances of a difficult range of mountains, deep rivers, with the possibility of Indians, and one cannot but admire the self-reliance and courage of these men, which can only be partially appreciated until a visit has been paid to these silent prairies. Anyhow it is a peculiar display of those qualities which must die out for want of use, so soon as the vast uninhabited areas are by degrees dotted with habitations. While out in camp one is cut off from the world. With elaborate arrangements, you may be able to receive letters at various small post-offices, once a month; your letters to the world may be sent oftener by the hand of some passer-by; but besides the disinclination to write in camp, due to a brain dulled by want of employment, and to the discomfort of scribing without tables or chairs, there is a feeling that you are not of the world, and your daily doings are not likely to interest it. Whoever is sent as messenger in to a store or town is bound to bring out some newspapers; and after the gap of a month you are startled with the conclusion of some event which, in the language of the broadsheet, shook the world, but of which no tremulous feeling reached the little camp in the dell.

The remoteness of other humans, and the charmed solitude of your temporary home, wraps you in selfishness; you are glad of your isolation. Everything around you is beautiful. Far off the ranges lie one behind the other in fading tints of gray and blue; halfway the broken ground of the bad lands assume fantastic semblances of long lines of walls and towers; the red stone exposed in the steep scarps, glowing in the bright sun, is banded by sharply cut black shadows; one bright green patch shows where the creek spills its water in the spring, which has given birth to some acres of meadow. The whole land is silent, desert, and a bit mournful ; it seems to be slumbering in the haze, a long sleep which began at the creation, and is waiting for a call to life. The nearer hillsides are colored by acres of lupin and sunflowers, and all over the valley there are various flowers, brilliant but odorless ; higher up in the hills are wild roses, wild currants, and gooseberries ; the latter are particularly good, and will temporarily sweeten a life supported too exclusively on the dry fare of the prairie. The grass is rich and thick. The cattle having retired up here, out of reach of mosquitoes, are happy all the day long; they eat in the cool, and lie down in the warm sunshine ; look after the calves, and grow fat as seals : if only they are not endowed with knowledge, and cannot anticipate the visit to Chicago, from which there is no return, or the bleak winter, with its short rations and piercing storms. The life of the cows is not all peaceful ; reasonable family happiness includes a changeable amount of quarrelling; and though the bulls generally live apart and in harmony, there must be some rows. A peculiarity was connected with these events, that the business seemed to be adjusted at certain fixed spots, a flat clear place being selected where an assembly was constantly seen ; here the strangest evolutions were gone through, which, watched from a distance, it was impossible to understand. It may have been a fight, or a medicine dance, or a ceremony. The steers would be excited by the first appearance of a small gathering, and scream and run to the ground. Here they would circle round the centre knot, who, by the antics of pawing up dust with their fore-feet, were probably bulls about to settle a dispute ; the cows would stand about outside and graze quietly ; dust would rise in clouds. After a little time the hubbub would quiet down ; there was no fight ; perhaps it was a meeting of the house where personal views had been exchanged, much as carried out in the parliaments of crows and other bipeds. In the spring the cattle sometimes get hold of some poisonous weed ; this is said to be the root of the lupin, which is one of the first plants green, and is liable to be snatched up by a greedy cow. The effect is to make the animal swell enormously ; it then becomes giddy, and in a few minutes, if not relieved, it dies. The only remedy ever tried is a stab in the side, which, in the one case I saw applied, was not successful—the cow died.

Our camp was near one of the trails which, not so many years ago, was the only road into Western Montana. The middle portions of the Yellowstone Valley were the last country in the Northwest to be explored. and had then been penetrated by few, a wholesome regard for Indians, Sioux, Crows, and others, deterring any but the boldest and most adventurous. This trail was still used by the Crows, who came down to visit the Rapahoes, Shoshones, and other tribes, on a reservation round Fort Washaki. If only braves in the company they were seldom in greater numbers than four to six, but they were always disagreeable visitors, wanting to be fed, but too idle and dignified to assist. Their knowledge of English was limited to one or two coarse expressions, which are jerked out with a malapropos highly diverting. As a rule they are silent, and give a grunt, or make a sign, to avoid the labor of speech. It is a great pity that nothing can be done to save their sign language, which is sufficiently elaborate, and is capable of communicating all they wish to say or learn in their simple existence. Two Indians will converse for hours without a spoken word, and tell each other of the events which occurred during the lapse of time since they last met. Having seen Mongolians in the Gobi desert, I could not help noticing, both in appearance and habits, so far as their outward life is concerned, a number of small resemblances between the Red ·Indians I came across and the Mongolians. The skin of the Indian is yellow, but when abroad he covers his whole face with a red powder; it is said to prevent sun-burning. The young fellows and women often rouge their cheeks only. The Crows are a tall race, given to brilliant blankets and Jim Crow head-dresses. They are not at present troublesome; in fact, they rather look to the white man for protection against some of the nearer tribes, who rob their horses, on whom they have not the courage to retaliate.

People who know nothing about Indians look at them at first with curiosity, which soon is mixed with a little contempt; but those who have had much to do with them in wars dislike their presence, and, knowing their habits, are often nervous and apprehensive of treachery. It would be a meritorious deed, from an Indian point of view, for a band to murder a single white man, if it could be done with perfect safety in regard to their own skins, and to the business getting known to the agency, and being visited upon the tribe. Their silent, stealthy ways always look suspicious; and if, seated round the camp-fire at night, you hear a gruff "How!" like a bark just behind your back, when you thought no one within fifty miles, it makes the party jump. A red man has walked silently up, and is now standing nearly within the circle. Few of them are bashful, so he asks for something to eat at once. You let your visitor have whatever scraps were left over, and if the grouts have not been thrown out of the pot, water is poured on, and the liquid set to boil; this second decoction of the berry, much resented by the tardy cow-boy, goes by the name of Indian coffee. While the Indian is thus occupied, the party are canvassing the possibility of his having come in as a spy, the rest of his tribesmen being concealed not far away; what devilment they are up to; whether they will run off the horses in the morning, or rob any part of the outfit while we are sleeping. It is, however, of little use our troubling ourselves in anticipation. Our horses could not be found in the dark; we cannot discover anything of the Indians' numbers or intentions before daylight. When morning comes, it is generally all right; the Indian is alone. Having breakfasted, he tells us that he will be travelling south five days; that is, he points in the direction and says, "Five sleep"; he shows, with his right hand passing thrice over the back of the left, which is held knuckle upwards in the shape of a hill, that he has to cross three ranges; at the end he will camp, which he denotes by closing his fist, and making a jerk downwards as if pushing in a peg. We speed the departing guest, and return to business better satisfied for his going. The possibility of the Indian being converted to any civilized or useful purposes is a chimera;

he will be a wild man, or he will die out; his inherited disposition will prevent his ever being a satisfactory member of a settled community. On the frontier a good Indian means a "dead Indian." Whether the Indians have deserved, or brought on themselves, the injuries they have suffered, and to what extent their treatment might have been ameliorated by honesty in the agents employed by the Government, and by a more humanitarian spirit in the people who have ousted them, can matter little at present. The Indian must go, is going, and will soon be gone. It is his luck.

About the middle of July, it was time to leave our summer quarters, and make for the Yellowstone country. The trail we were following certainly deserved no better name; there were tracks of other wagons having gone by that route, but it was uncommonly rough work. The way down from the summit was along the backbone of a long ridge, so narrow at top, with sides shelving down so steeply, that from a high driving-seat in places you could [look into [the bottom of a valley on either side, without much appearance of the intermediate ground. After the first descent of the mountain-sides by large, bold slopes, the foot-hills consisted of a network of small ridges and valleys, without trees or grass apparent. Looking down on this chaos, the tops of the broken ground seemed all one level; there was not a feature of any salience. To cross it would be a constant up and down, without shade, shelter, or water. If there were cattle or wild animals in these sterile regions we could not see them; probably there were not any; and that it is one of the pieces of country which will not come into use without the machinery of a subsidence or an upheaval. The Wind River, after leaving the Shoshone reservation, enters a canyon, and comes out under the new name of the Big Horn; it passes through another canyon before it joins the Yellowstone. The ford is not far below the first passage. In July the water is not very deep, but much before that month the river is not passable by carriages; not that there is much of that kind of traffic except the cook's wagons, which follow the round-up. While halted on the banks a band of cattle were about to cross. These had been a day without sufficient water; therefore, so soon as they were aware of the river, they began to step out. The leaders, long-legged and stout steers, got away and travelled in splendid form; the smell of the water sent them bellowing; the dust of some two thousand animals, following in a long file along the trail in a parched country, formed a rolling, heavy cloud in which at a little distance the cattle were themselves concealed. The leaders of the herd were getting close; they disregarded the house and our small party on foot, intent only on reaching the water. Their necks are outstretched and mouths open with continual bawling. It looks like a walking race; they shuffle along with the most busy determination, careless of aught but the getting over the ground. One steer steps out of the line, looks back, and gives a long bellow, as if for one moment he remembered his chum left behind; but he turns again still more quickly, rejoins the line, and travels to make up time. At last they reach the river-bank, which is not very high and has been sloped by the hoofs of former herds. They rush right in till the water is half-way up their sides, and then settle down to enjoyment. The remainder follow quickly. All go in, drink, wallow, and stand round; a few climb out on the further bank, full of water, content, and too heavy to wander. The last division comes up, of cows with young calves; a troublesome lot, always trying to stop or to get back, who tax all the patience and energy of the boys left in rear to bring them along and compel them to rejoin the herd each night. These last are driven into the water, and, when all have had enough, the herd is taken out and collected on the far bank.

The wagon having completely smashed up as a consequence of the three days' journey down the mountains, a selection of properties has to be made and luggage has to be reduced to smallest limits; it is packed on horses, and we go forward much less encumbered and at a considerably better pace. Except for its novelty the country in the Big Horn valley is not interesting; the river bottom is filled with thick timber, but on the low hills there are nothing but scrub-bushes. A few buffaloes had wandered into these regions; they were generally two or three together and rather wild. They gallop with a clumsy, lumbering gait, as if too heavy in front; the head, carried low, nods, and the shaggy fringes of hair on the fore-arm flop with each stride; they are at their best over broken ground, and drop and disappear into a deep ravine as if through a trap-door. We found water every ten miles or so, and the horses jogging along under the burning sun would be glad of these chances. The troubles of saddling, arranging the kit, lashing, tightening, keeping a continual watch on the loads to anticipate a turn-over, to catch the horse whose pack is tottering, and either take in the slack or do the lashing over again—these are the hourly interests in travelling with pack-animals. The horses run loose, and are driven in front; they learn to keep the trail. If they break out to one side they must be turned in again. After the first half-hour all the ropes must be overhauled. One man heads and halts the herd; they at once split up, wander apart, and feed. You must dismount and catch them one by one, to secure their loads; but many of the horses object to being caught, dodge us on foot, and run. If there is a third man he remains mounted and turns the fugitive; but if possible it is best to avoid exciting the pack-horses, for when they gallop the load, if somewhat loose, falls to pieces; the saddle perhaps slips and the horse finds sixty or seventy pounds, weight of blankets, grub, etc., under his belly, with a tin pot rattling about his legs. This is too much for his nerves—a few good kicks, and he is rid of the whole; jumping clear of the tangle of rope, he is a free horse, and leaves scattered on the prairie the contents of his pack. Before he can be loaded again he has to be caught, which means a deal of galloping among the ordinary experts at lassoing. By the time the saddle and pack are restored, three-quarters of an hour have been wasted. Viewing the trouble of loading, the daily journey is done at one stretch. Arrived at the halting-ground the ropes are thrown off, the bundles taken down, the saddles ungirthed, and the horses turned loose to roll, drink, and feed. Only one horse is kept tethered.

Some travellers tether all their horses; but this requires very good feed and clear ground, otherwise the horse starves, either from scarcity of grass within reach, or by catching his rope in the bushes he is reduced to the circumference of a few feet. Many horses will start off by this manœuvre: they walk round and round the first stem the rope catches against; having thus wound themselves up till their noses are in the bush without a chance of anything to eat, they remain thus foolishly prisoners until relieved. To hobble the horse is an intermediate plan, which also requires the feed to be tolerably thick. The hobble is a leather strap which joins the two fore-feet. If not holding them very closely together, the horse has sufficient freedom; he can indeed, if he wishes, get away a good distance. The objection to hobbling is that the horse may be thrown and injured, and if grass is scarce he does not get so much to eat as if entirely at liberty. A loose rope is sometimes left hanging to the necks of one or two; this leaves a good trail to follow, and checks their wandering somewhat; but it also has the disadvantage of holding in the brush, when the horse may starve, or the hind-foot may catch in the loop round the neck when the horse scratches himself. To meet this the noose is often not checked by any knot, so that it will give; but then it may also draw tight and throttle. The best plan is to turn the horses quite free to go where they like and forage for themselves. In this way they will get most rest, the best feed, and remain in the best condition for work. The band also hangs together, and sometimes are so far well trained that they will not stray a great distance from camp; but with the usual half-wild horses there is no certainty,—they must be watched, and if they have already wandered far they may be driven back to near camp just at dusk. This is rarely of use, for so soon as they are left the horses are apt to turn and go straight back to the place they have fancied. There are often in a band one or two inveterate strayers, who will lead all the others off. Some recollection of grass or water at a former camp, or an idea of returning to the home-range, or some other pernicious idea, gets into their brain; they set their heads in the direction and travel; not at any great pace, but quietly foraging and walking. By morning they are ten to fifteen miles away; in this disposition should they strike a trail they will follow it, one after the other, with little thought of feeding, as if they had travel on the brain; no doubt accomplishing a fancied duty or falling into an old habit.

Before dawn, the man who has to hunt the horses must get up and saddle the one which has been picketed; he rouses the cook when starting, but leaves the others snugly rolled up in their blankets. Riding in the direction where the herd was seen the night before, he takes up their trail and follows that till he finds the herd. They seldom have gone further than a mile or two away, if feeding properly. A hazy morning, or rolling country, makes it a little more difficult to find the herd. The safest plan is to stick to the tracks; if one or two of the horses are shod, their footmarks are plainer, and satisfy you when you see them that you are not following the trail of a strange band of range-horses. The horse you are riding will often catch the scent of his fellows before you see them yourself, and lead you right up to the herd feeding in some sheltered hollow. You count them, get behind them, and, with a hollow, start them at a trot towards camp. They know very well what is required, and, if held together in a single mob, will not try to feed, but jog quietly along. Now and again one will free himself from the ruck; you will find that your mount will go straight for him. A knowing horse will do the driving entirely by himself with little directing. On nearing camp the men there have made a corral by tying two long ropes to a tree; a man at the end of each holds the rope taut, and into this V the herd is driven; the man on horseback watching the mouth, the men on the rope checking by a jerk any attempt to break out at the sides. The horses required for riding or packing are caught up and tied, the others are let go to find feed about until time to start. If travelling alone, with only one or two pack-horses, they are often led instead of being driven in front. This is not so convenient for watching the loads, but has to be done if the horses have not been accustomed to follow a trail of themselves; otherwise they would continually diverge, and have to be driven back, which would lead to galloping and overturning of loads. With two or three men it is better to drive. But, after all, packing takes from the pleasure of your trip; it should only be practised where no fashion of wheeled vehicle could possibly come through. Except high up in the mountains there are not many such places to be found. A tough wagon, a moderate load, four good horses, and a skilled driver seem to be able in the West to go anywhere, or to get round, which amounts to the same.

After several days' travel we reached Clark's Fork, and took up our abode on the slope of the mountains.

On the further side these mountains flank the eastern limits of the Yellowstone Park. There are two ways of entering this from the east: one directly, a trail by which General Sheridan's party came out. The other route goes a good deal south. Once in camp, a tough job stares me in the face—that is, to wash my clothes. It is a struggle to keep a small quantity of water hot; the only tub is a box, which should not leak. A few under-garments of flannel occupied me a long day; for, after washing, the articles were so full of soap that no amount of rinsing seemed to take out the clammy feel. The soap used in the West is a strong chemical one, admirable in washing up tin-plates and greasy dishes, but I was afraid of what might result to my skin if I dried the garments and wore them. Warm water having played out, I dipped and wrung, and dipped again in cold water till my hands and arms were numb. My coat was off for better exertion, and the wind blew chill through my shirt. Hopeless of success, I give up work, lay the clothes in the stream, fix them with stones, and leave them till next morning. An aching back checked any notion of further washing, so the bushes are festooned with articles, gray and white, where they are left to dry, and, what was not necessary—contract. One or two washings satisfied me that it was not my line, and in future I eke out my wardrobe, if possible, to reach some village where a Chinaman was to be found. These fellows cheat you horribly; scamp the work, and charge you at the rate of four shillings for six or eight pieces; but even their most inferior and overpaid work is preferable to the best results of one's own exertions.

The climate and scenery were superb; on the mountain-sides miles of forest, dreadfully ravaged in places by fire; at the foot miles of grazing. At this time all quite unoccupied, as we are on the edge of the Crow Reservation; the only way for a white man to settle in an Indian reservation is by marrying an Indian wife. This, from all accounts, is a one-sided affair. The squaw-man, as he is called, always looks ashamed of his weakness. He is very much married; she can be divorced by Indian law with great facility. His friends are ashamed of and avoid him; her friends and relations come and live with her and on him. On separation the property is hers. He has nothing. What then attracts a man into this connubial state? It is hard to say. Probably an idle, vagabond life of trapping has led to continual association with a tribe and a tolerance of their squalid and dirty habits. The memory of a better condition fades, and after a time an uneasy dread of returning to old associations drives him into braving the alternative. There are plenty of good streams flowing out of these mountains, in which trout of all sizes and grayling are caught, the ordinary bait being a grasshopper. Not much skill is required, and anybody can land a dozen or twenty of about one to two pounds in weight, in a few hours' fishing. Where a stream has not been troubled as many may be caught in an hour. They will rise to an artificial fly, but the grasshopper is more certain, and when fishing for the pot the latter is preferable. The rod is often a willow, cut on the spot; the line a thin twine tied to a hook. Nothwithstanding the uneducated simplicity and greediness of the trout, skill has much to say in filling the basket. One man after reaching camp in the evening will bring in enough for supper; the rest of the Waltonians not catching more than a fish apiece. One thing sure—we can all eat them. There are some pink-fleshed, some of a more yellow tinge, and many white. The small dark-skinned brook-trout are particularly good; they are picked out of the pan with unconcealed selfishness. The bigger ones follow; they are all good enough to people who have not tasted fish for six months. I am ashamed to think of the numbers fried and set down for our meal, and which I helped to make disappear.

Having a little time to spare, I made a trip round by Prior's Mountain, and through Prior's Gap. This pass is from a quarter to half a mile broad, and about ten miles long; it is on the road from Stinking Water to the Yellowstone. The sides are granite crags boldly scarped, which shut in the pass on either side; a stream runs through a great portion of it, and the bottom is splendid grass. The rocks are of very rugged outlines. A road will probably run through this gap hereafter; it will certainly be one of the most picturesque in this part of the country. After coming out of the gap, we follow down the creek, which we are loth to leave, as it is a larder of most excellent trout; the road eventually brings us to the Yellowstone at Coulson. This is a little place which made a beginning before the railway was started, its trade being the supply of food to the emigrants around, and the purchase of hides, etc. With the arrival of the N. P. Railway, Coulson was at once smashed, on the traditional system of these railways. Subsidized by land grants, which, if anything were ever made known concerning these railways, might be found to have covered the first cost of construction, the railways may in this way be said to have been constructed at the expense of the nation; in return, they have been laid out with the narrowest ideas as to suiting public convenience. If including a large town in their scheme which cannot be avoided, the railway depot is located at a mile or a mile-and-a-half from its main streets; the intervening land is taken up by the company, laid out in building-plots, and sold to speculators. If a small town is approached, the railway company sets up a rival close beside it, with the intention of crushing it. They erect a few sheds, and maintain a few work-people; the land round is laid out in streets, the railway officials call it after the name of some local magnate, and a fair start is thus given to Snooksville or Pogram City. Promises are made that a round-house for engines, workshops, etc., will be built; a hotel is subsidized by one or two trains being halted for passengers' meals; saloons spring up to meet the demand of the workmen for such delights; the store is transferred from the earlier rival on account of the convenience of being near the goods-shed; in a few months the business is done—the railway town prevails. In this way Coulson had diminished, while Billings had flourished more like a gourd in a night than a green bay-tree; it was but three months old, but as the Americans say was quite a town. Facing the railway, on one side, was a line of plank buildings—hotels, stores, and saloons galore, with strange names, e.g. the Bank, the Exchange, the Pet, the Cozy. On the other side more buildings again, mostly saloons, are being run up; to the back private houses are started, of no great size, but good for a commencement. Billings booms! Foolish prices are quoted for building-lots; they may be paid. Banks and stores do some business, but the main trade of the town is at the bars of the saloons or at the tables of stud-poker. Of all kinds of hard labor, keeping a saloon is most to the taste of the ordinary citizen. The stock-in-trade is small; a barrel of whiskey and popularity are the essentials. The saloon may be a single-roomed plank cabin, neatly papered. On the walls may hang pictures of Abraham Lincoln and General Garfield, with a few comic sporting prints. A bar runs part of the way up the room, and is spotlessly clean; behind this counter against the wall are a few shelves decorated with specimen-bottles of wine, spirits, etc.; underneath, sugar, lemons, and ice, if these luxuries are attainable; a stove, three or four chairs, a bucket of water with a dipper, complete the furniture. The saloon-keeper is always tidily dressed, appears in a white shirt, his sleeves and wristbands protected by calico cuffs; his cleanliness, and his not wearing a hat, at once separate him from his customers. With these he must maintain pleasant terms; receive and retail news social and political; serve them when they wish to drink; scrupulously wipe dry any slops on the counter; and keep the stove supplied with wood. In these he fulfils the whole duties of a saloon-keeper. He seldom moves out of doors or from behind his bar, unless, indeed, in the case of a row, and the boys begin to shoot. It is wisest then to withdraw as quickly as a happy thought, returning to the daily routine when matters have quieted down, or to prepare for the rousing business which follows with the coroner's jury. Land offices offer town-lots in Billings at all prices, and, to farming settlers, quarter-sections in the vicinity, to be irrigated by a ditch, which will be led out of the Yellowstone twenty miles above the town. So long as the railway keeps Billings as its terminus it continues to improve; but when the railway opens out another length, the prosperity which flows from the expenditure during its construction moves off again to the new terminus. The railway have by this time parted with their interest in the land; one guesses the result. Billings, however, has better chances than other small bantlings of these railroads. The hotel and restaurant are wonderfully good, considering. The first is crowded; your bed is a dollar; meals, 75 cents each; 50 cents if a card of twenty-one are taken,—not too dear. You are fortunate in getting a bed to yourself. In these frontier towns you are liable to be told off to share a bed with any stranger; a process which was only avoided by much manœuvring, and by accepting in a spirit of content any other accommodation. Stabling your two horses over the way costs two-and-a-half dollars, as much as your own daily bill; all goods are very expensive, except whiskey, which is at the ordinary price.

The effect of this high living on an economy trained to lower fare is disastrous; in two days I am feeling so ill that I pack up and start out again to rejoin camp. My companions for the next few days are two hunters; one a Frenchman, the other was an American, and rather typical of the country and trade. Judging from subsequent experience, I should say that the profession of trapper or hunter no longer pays: beavers are scarce and buffaloes dying out; their skins were the staple products of the chase. A well-known hunter used always to be able, at the commencement of winter, to get his outfit and food advanced to him by any storekeeper. When he brought in his hides in the spring there were sure to be enough to clear the debt and leave a handsome surplus. What became of this surplus may be guessed, when next autumn found our hero broke and again depending on advances. The race had its traditions. To do anything with money but spend it recklessly was below the dignity of a buffalo-hunter; as one man put it, referring to the saloon-keepers and gamblers, who are invariably the neatest dressed individuals in the crowd, "Some one must support these white-shirted sons of ——, and I should like to feel that I had done my share." My American was very shrewd, and had evidently talked much on the inequalities of social life elsewhere than on the prairie. He often posed questions on the habits of the old country, which I could not always answer satisfactorily even to myself. A trick these Western men have, when finding fault with men's ways in civilized places, is to ignore the life of the greater mass of their own countrymen, and to illustrate the injustice of man to man by the practice in Europe, naturally, of course, particularizing England, being the country of which they have read most. Can a man in your country put his gun on his shoulder and go for a hunt? Can a man homestead 160 acres? No! A Western hunter naturally considers a Government which does not supply these outlets to the industry of hunters as nothing better than a despotism of the upper classes. But the Americans in the East, I point out, cannot shoot over their neighbors' lands; the trout-streams are all preserved; and if a man wants a farm he must buy it. But the American can always come West and settle, is his satisfactory reply; an alternative bound, in his view, to last all time.

Western men will carefully say ladies, in speaking of the women of the laboring classes; but concerning those of better education, refinement, and circumstances, woman is good enough; it is not a matter of chivalry or rudeness, but simply self-assertion. The exclusion of servants from the use of the ordinary sitting-rooms seemed also to be a wrong. If a man had washed his face and hands there was nothing inherently defective which would unfit him from sitting at table with any society. As to servants, when the day's work was done, and they had dressed themselves, why should they not sit in the same room with their masters? It could do the latter no harm; whereas, if the masters were superior, the servants might derive benefit from their companionship. Forasmuch as all men are equal, the fact of one having more money than another could not unfit them for each others' society. On this occasion I must have been routed, as after a few months' life in the West my ideas on the advantages of different classes and social layers had been lying torpid, and were undermined by the general habits of equality, which were in no way irksome among men who lived simply, and were occupied entirely in the open air, and who could therefore live together without the close contact which would be felt if they were confined within the limits of four walls. We remained good friends, but unconvinced, and so kept our arguments for further discussion and mutual improvement. I was much flattered at his surprise, when I confessed how short a time I had spent in the country. My advanced knowledge of packing was the quality which most recommended me. Unfortunately, with all our talent, there was no game to practise on; the mountains were full of elk, bear, deer, etc., but the cold had not yet driven them down. In the valley there was nothing except antelope, which were wild; the river, however, had plenty of good trout, though it was getting a little too late to kill them. Having just left a town, our larder even of luxuries—potatoes and onions, I don't well remember anything else—still held out; we did not make very long journeys, and the weather in October was all it should be.

While travelling south of the Yellowstone, we had been all the time in the Crow Reservation, and saw many of that tribe, generally living in their "tipis" on the banks of a stream. The "tipis" are the conical tents, which are supported on a bundle of light poles held together by a lashing above, while the lower ends are stretched out to leave a good circular space within. The covering used to be made of skins, it is now almost always of canvas. The cut of cloth is very simple; when laid flat on the ground the covering is a semicircle. There is a little arrangement by which, as erected, a sort of cowl protects the orifice at the top and prevents the wind blowing the smoke back into the interior. The edge is held down with pegs, but in ancient days, before tools were invented or in common use, large stones were laid on the skirt—at least this is the explanation of circles of stone which are often found on the prairie now partially buried. It is a very commodious and comfortable pattern of tent; the principal trouble is the number of long thin poles. In travelling, these are tied at their upper end, and slung over the back of a pony; the lower ends trail on the ground, and tell very plainly of the passage of an Indian family.

The Indians constantly visited our camp, and were very impudent and persistent beggars; our main dread was lest they should steal cups, or knives, or articles we could not easily replace. One morning we missed a bag holding all our tin-plate; we of course suspected a batch of women, who had held out by the camp-fire over night until everything had been washed up and packed away. By signs we complained to the chief, and produced a general hubbub in their encampment of dogs, children, and chattering women. Tne chief, to assist us, ordered out his forces, and made a line of some ancient and smoke-grimed crones; in this formation they beat the bushes round our camp and soon found the bag, which had evidently been carried out of camp by a dog or cayote, who had been attracted by a piece of jerked meat within. Our faces were accordingly covered with shame for our wrongful accusation, and the old ladies let us have an expression of their views on our conduct. We immediately distributed biscuits and coffee, and made peace. It was abominable on our part to suspect them; but if the nobility of the land enjoy a doubtful character, are dirty, unkempt, bundled in rags, and sit round your camp-fire chattering after you have all gone to bed, the suspicion naturally follows. They are themselves apparently careless of property, and in a camping-ground recently occupied by Indians there are often several articles of dress or the kitchen left behind. On the road the care of everything is handed over to the women, who pack the tents and luggage on the miserable ponies—mostly thin, a few crippled. It is a great sight to watch the outfit strung out on the trail. During cold weather the very old woman in the family will be wrapped in a buffalo-robe, which wrinkles and bulges in hard angles, making her look like a gray rock balanced on a pony; the children are well-folded in blankets and hold on in pairs or singly; the young and middle-aged women do the driving, and keep the herd together. They wear striped blankets, and of course straddle their horses; but it is a most disorderly rabble—one pony here, another there; colts running about losing their mothers by staying back, and galloping wildly into the hind-quarters of two or three pack-ponies before they tumble into the right place. The pony dragging the tent-poles is out in the bushes by himself, fifty yards from the trail; he will probably brush off a pole or two. A girl, apparently five years old, riding a beast with three straight legs and a crooked one, and with only a string-halter, is drumming her little feet and whacking her mount with a willow wand to make it leave the road to hunt in this stray. Two little mites are perched together on a mare, which stands still to let its foal suck. The whole crew is scattered along a quarter-mile of road; at the tail ride leisurely the two women who bring up stragglers, and keep the business moving. So soon as they are near, the cavalcade closes up, trots on a bit, and in half a mile are scattered as before. But where is the brave? He knows of some encampment off the road, and is paying a visit, eating bread, meat, and drinking coffee—a duty more to his taste and dignity than looking after squaws and children. The Indians look contented, and I suppose get enough to eat; but the food allowed by the Government is said to be insufficient, and the Indians are expected to supplement it by hunting for themselves. A number may always be seen round slaughter-houses; the butchers contemptuously throw the guts to these poor wretches, who seize and eat them raw in a disgusting manner.

On return to camp I amuse myself for a few days in fishing, and then join the boys, who are taking a herd of some three hundred steers to the railway for shipment to Chicago. There was no particular hurry, and though the country was stony the grass was very good. The herd might travel leisurely northwards and add flesh as they moved. We had the country much to ourselves. The weather at first was steady, and generally bright; the cattle lay still all night, and were found together in the morning. They travelled a few miles every day. Water was plentiful. We had no difficulty in camping and always had plenty of wood. Matters went smoothly and pleasantly for a time; but when a change of weather set in, and the rain was accompanied by snow and severe gusts of wind, the cattle commenced to break up and to look for shelter, wandering off a good distance, so that each morning the count was short and the strays had to be hunted. It was hard work for the men and the horses; for, once started, the cow-boy who respects his character is bound to stay with it till he finds the cattle he is hunting. The pace must be good, as the quicker you come up to your strays the less distance will you have to bring them back. The same cold storms that troubled us harassed the wild beasts on the mountains, and the bears came down, following the cover which fringed the streams. The scent of the bears to cattle and horses unaccustomed to such neighbors was terrifying, and both began to show a strong disinclination to go through the bushes and trees to water. Our most troublesome customers were half-a-dozen bulls. They never cared about keeping with the herd, and each day all or some would disappear, generally in pairs, choosing the heaviest cover in which to hide themselves. We had no tent, but managed pretty well. Several mornings I found the canvas cover, which I had pulled over my head, cold and stiff. Having pushed it back the scene was very pretty—the trees and ground covered with snow and hoar-frost; but it did not look inviting to turn out. One by one other heads would appear, take in the landscape, and not like it. Then the cook made his effort, and having started a fire we would creep out and help pile on the logs—a business in which the laziest will learn industry; with the added hope of speeding breakfast on a frosty morning. Our fire-wood was generally of birch trees, which had been water-logged, and were decayed just above the roots. A pull would bring down a bare pole, thirty feet high, and with very little trouble in chopping one collected a good pile of logs. The unfortunate horse which had to remain saddled and tethered all night was a sorry sight; his feet drawn together like a goat; his coat staring, cold to the marrow of his bones. How cheering the sun was when it did show itself! The party distributed—some to look up cattle—two to catch the ponies and pack the kit, a job none the more pleasant with ropes wet and frozen, hard to put on, bound to stretch loose.

I made a little independent trip to the Agency, partly to visit it, and in part to buy some food. The place was only interesting from the Indians congregated there in sufficient numbers to enable one to get a sight of some of their ordinary habits. Their "tipis" were generally set up two or three together, the interval between them being sheltered by an erection of poles carrying a flat roof of branches and leaves. Music issued out of one of these homes. On looking I saw a circle of some thirty men and women—the two sexes sitting apart. At the side furthest from the doorway sat three or four chiefs, with strange bundles lying in front; two men facing them, standing in the middle of the tent, held otter-skins in their hands, and were dancing—or rather hopping—alternately on each foot. Men were singing, drums beating, and a skirl from the women came in at regular intervals. It was a medicine dance, for the purpose of increasing the virtue of the chief's medicine. The scene was certainly interesting; the music, though monotonous, was not disagreeable. The only other ceremony I had seen was a buffalo dance, performed by a party setting out on a hunt. A charmed circle had been erected of green boughs, and within this a dozen braves, most fantastically dressed, danced round the enclosure to the accompaniment of a drum. It was hard work, and after a few minutes all would sit down and take breath. Some of the actors appeared to imitate the motions of animals, but the ceremonies in both cases were difficult to understand. On my return I succeeded in missing the trail, and had to camp out under a bush with two thin horse-blankets for bedding. I made a good fire, but the necessity of feeding it through the night rather interrupted sleep. Both horses were frightened by bears and broke away. Next morning I started early on foot for camp, where I was deservedly laughed at for my clumsiness. But I got breakfast, which is a comforter. I found one horse, and borrowing another in camp went back for the luggage. The other horse was caught the same evening by a piece of good luck: I was to have sought him the next morning, and had intended to hunt for him in quite the opposite direction to which he was found. One of the boys had shot an elk; the agent had presented us with beef, potatoes and vegetables; the stream was full of trout; we were in clover. We started for Billings, and after several days' travel, with the alternate fortunes of losing half the cattle or losing them all during the night, the boys thought it best to watch them.

In this way all were virtually brought to the post. One big steer was bogged in a quicksand, and could not be pulled out with the horses. We put ropes on his horns and on his tail, but by neither extremity would he be extricated; an Indian offered to dig him out, which he did. He first sent to his home for assistance, and three old women came down on a pony. They made, with long grass and mud, a dam round the steer, and dug him out till nearly free, and then got him ashore; the next morning, however, they cut his throat to prevent his dying. Another steer, in falling, knocked off a horn; the smell of the blood sent the whole herd wild, and most of the steers occupied themselves in hunting the wounded one up and down the line without cessation. A donkey with her colt was one of our embarrassments; they had a continual foolish desire to join the steers. Although these had seen the pair constantly, they never grew reconciled to their novel advances, and so soon as the four long black ears loomed above a bank, up jumped the herd and made off. A vast amount of strong language was hurled at the offenders; of this they recked little; while to drive the pair away was not so easy. The boys on horseback had nothing handy but a short whip hung on the wrist, while the jinny had her heels, and was expert in the art of self-defence. These pranks of the steers were not attention to business, which would have rather been an orderly advance, eating grass, drinking water, and laying on fat; but the beasts were altogether in a bad frame of mind, and annoyed us by bawling and snatching the heads off weeds and brambles, showing they were discontented, although in the midst of plenty. It is curious what trifles will disturb the cattle, and the man managing a drive must always be considering how the steers will be least discomposed. The horses' manes and tails, at this period, were stuck full of burrs; the forelock thus twisted and knotted became a solid pad, and the tail moved in one lump. One horse had an awkward fall while galloping through high grass. He put his fore-leg into a hole left by the burnt-out stump of a tree, which had smouldered down to its roots, leaving a cylindrical pit, about a foot wide and four feet deep. The rider caught under his horse in such an accident has a chance of serious harm from the projecting horn of the saddle, should the horse roll or flounder much in struggling to his feet. These holes are very common, and, as they are difficult to see, a man on foot even is liable to an ugly fall while pushing his way through the jungle.

After fording the Yellowstone, the stock were put on board the cars and sent off to Chicago. The N. P. Railway was not in very good order, having but recently opened to traffic a great portion of its line. We met with two accidents. In one three cars ran off the road; the first was turned over on to its roof and burst open. When we recovered the jerk, and looked out, six steers were on the horizon, in good line, nodding their heads, and walking with the best foot foremost; they had had enough of the game. Another time we collided at a cross-over, and the guard's caboose had the end knocked off; there may have been minor events, but they were not brought to our notice. I should think that the loss in damaged rolling-stock was very large on these new lines. A considerable part might be prevented by a little more care in laying the way before it is opened to traffic; but we all fancy we know other people's business better than those concerned. In the West, the ruling idea is to spend the least time and capital. The American does not seem to care about a work being finished. The eaves of a house are often not sawn off to a line, or the planks on a bridge are left jutting out on both sides in a ragged edge. "It's good enough; it don't hurt the bridge."

After six or seven days' travelling we reach Chicago, and turn the steers into the stock-yard. The mass of cattle, sheep, and pigs assembled in this area is of itself extraordinary; the buyers ride round, have a few words with the commission agents, and the business is concluded. By means of gates the number is counted, or a selection is made and driven to the scales; for the price paid is so much per pound. This scale is in a building, the floor a weighing-platform, to hold, I think, about 200 head of large cattle, say, running about 1400 lbs. each. So soon as they are all on, the gates are closed, and in less than half-a-minute the weight is ascertained on a bar in an adjoining room. This is recorded, the front gates are opened, and the cattle stream out eight or ten abreast. Their number is at once called out to the clerk. They said that during the process of weighing, which altogether occupies but very few minutes, the man inside counts the herd twice; once as it stands on the platform, and again as it dashes out after the gates have been opened. With small Texas cattle, the number weighed at one process is over 300, and with sheep and pigs twice as great again. I can only answer for the cattle being counted, though I have no doubt about the others. To be able to count so many was certainly surprising, for I have seen several cattle-men discuss for five minutes whether there were twenty or twenty-one head in a pen, nor could I myself ever count that number twice correctly.

Going in search of a hotel, we were met by many touts; it was amusing to find one's-self accosted with the patronizing air which the city man affects towards what he considers the wild man. One agent was most pertinacious, greeted us loftily as "boys," and insisted on our giving the preference to his lodging-house, extolling its comforts and table. Along the road he saluted several acquaintances as General, or Judge, and in an aside told us that they were living at his house. But our plans were laid with a view to compensation for our late discomforts, and having disguised my travelling clothes under a good overcoat, and kicked off my overalls at a corner when no one was looking, I trammed up town, and walked boldly into a good hotel. I often afterwards watched the desk-clerk hesitate somewhat at the dubious-looking visitors coming to his hotel. The weather-stained hat and clothes, somewhat dirty and spotted with smuts in the train journey, effectually disguised, in some cases, very solid men, and good company. In Chicago they are forgetting the traditions of the West; it is the clothes and not the man they now look to. This of itself transfers Chicago from the class of Western to Eastern cities; and when you have eaten a very good dinner, you will feel yourself at once in the van of civilization.

After a pleasant week spent in Chicago, I return to Montana, eventually making Miles City my headquarters. There I bought a horse, and packing a little bedding behind the saddle, set out on the 10th of November along the Deadwood road, to see the country, which is little settled, and is comprised in the buffalo range. I expected to reach a cabin every night, so, although winter had fairly set in, I did not take any food, nor the ordinary outfit which would have been necessary for camping, and would have required a second horse. The stages were long, each twenty-five to thirty miles, while one approached forty. Even on a fairly travelled trail, the beginner in prairie life is apt to go wrong, or to find his way with difficulty. The American is short of words when giving directions; you can seldom get more out of him than "Keep the main travelled road," or, "You cannot miss it; it goes straight ahead." You start out on a well-defined road, and outside the fence find one branch turns sharp to the right, the other, more used, goes straight on to the hills; an old hand knows at once that the latter is a wood-road for hauling lumber, and goes right. Say you go well for six or seven miles; the track, though not very clear, can be easily traced; the wheel-marks separate, but rejoin further on; but now there is a fork in the road—both are well travelled, which is it? The next ranch you know for certain is ten miles off on your road; not a living creature is in sight. You must make a choice, and perhaps not before a mile find out your mistake—you must either go back, or strike across country to regain your proper road. The off-road leads to some meadows where hay has lately been hauled; the most travelled tracks were decidedly along it. After travelling some time, instinct comes to your help; the trails are not then all difficult. Finger-posts are hardly known, and the few put up by private persons are shamefully misused; their general fate is to be cut down by some man too lazy to fetch firewood; they are often changed through malice or the local sense of humor. To misdirect persons was a common enough trick among ranchers. The accepted notion is that a man should ask no questions, but travel on his judgment; not a bad plan; but to insure a supper and shelter for the night is well worth a little inquiry.

The halt for the night was made at farm-houses or small stores, the accommodation a little rough, but good enough if you get a clean floor and the use of a couple of buffalo-robes. The horse was stabled, so there was no difficulty in the way of an early start, except the laziness of the people. If the men were away, the women were always for taking a holiday, and were an hour later in rising and preparing breakfast; they were even disinclined to get the usual staples ready, preferring to go without rather than work. On roads which are much travelled, the competition of the farmers who lodge travellers leads to quite another style of business, and then you can get your breakfast and start by daylight. At some of the places a traffic was carried on with the Indians, who brought in skins, sold them for money, with which they immediately bought food; the dollars were not five minutes out of the till. In one instance dollars were scarce, so that the Indian was content to receive pieces of wood, say seven for a buffalo-robe; he would then lay down two with the word sugar; two more, coffee, and so on, till he got rid of the encumbrance of money. The men and squaws have quite separate purses, and bring in their skins to trade independently. They are not so simple now but that they get a fair price for the pelts; but in their purchases they are imposed upon, or rather pay long prices. One good woman was very indignant at the way the Indians were taken advantage of at a neighboring ranch, and assured me that she gave them for their money *nearly* as much as to a white man. The journey from stage to stage occupied the best part of the day. Having found a man with a wagon going in my direction, I got him to carry the blankets. The day we started on the longest stage, we were overtaken by a storm, and camped ten miles short of the ranch; we could afford to do this, as his wagon had a good tilt and a bottoming of straw. The two mules and the horse were picketed by long ropes to the wheel, and we did the best we could for supper in gusts of wind and showers of snow. The apparition in the firelight of a strange white dog startled us. He was a civilized dog; we were suspicious; immediately thought of horse-thieves, and looked round at our animals. That night we lay in the wagon, a cold and disturbed attempt at sleeping; the wind threatening to turn us over, and the animals jerking their ropes as if scared; whereas, when quiet, we must need look out to assure ourselves that they had not been driven off.

The night, however, was too bad even to steal horses in, so no harm came to us. Next morning the wind lulled, but the weather hardened, snow still falling; but every indication of very severe cold to follow. We determine to halt at the cabin. At this solitary house, which had a room for the family, a half-underground cellar, and an unfinished large store-room, some eight of us were caught in a cold snap. The weather fell to over forty degrees below zero; and during the first day a sharp wind blew. The travellers kept dropping in; the first man had come from three miles down the valley; he started over-night, but feeling doubtful of his bearing, and blinded by the snow blowing, he found a hollow in the bank, sheltered himself and his horse, and lit a fire. His position had evidently been not quite comfortable nor reassuring. Then came in the post-stage from the west, the driver a little frost-bitten; he had come within half a mile of the house the night before, as a mitten he dropped there was picked up next day. In the snowstorm he got turned round, and went off seven miles on the back road, and had then stopped for the night. Next morning he was bent on pursuing the road in the same direction; but the passenger with him objected that the rising sun was behind them instead of in front. After some persuasion the driver turned round. Whiskey was at the bottom of a good deal of this trouble. A little before noon the boy came in riding with the post-bag from the other side. He fancied himself all right, and, though unable to face the wind with his carriage, had spent the night securely enough. But when we looked at him his face was discolored on the nose and cheeks with black patches of frost, and both ears were frost-bitten. The portion of the ear frost-bitten looked as if the flesh were turned into a yellowish-white marble. On going inside and taking off his boots, he found both feet had also been caught; the latter were dipped into cold water mixed with snow. The ears were lubricated with kerosene oil, and wrapped up in a hankerchief. I do not think much of this remedy. Cold water is the better plan; and, after the parts have been thoroughly thawed, oil no doubt is good to allay the surface irritation. The return of circulation in his ears seem to cause acute pain; whereas in his feet, of which at least so much as two toes and the joints were quite colorless, he did not complain of suffering. I often afterwards, during the winter, saw young fellows with blistered patches on their cheeks. With the heaviest of clothing and best of care it is impossible to avoid being frost-bitten if caught away from shelter during a cold "snap," that is, during the passage of a wave of cold, as newspapers put it. The snap lasts from three days to a week, as a rule; and the wise man, finding himself in shelter, stays there. We all stuck to our shanty, and fortunately had a good time. There were some excellent lumps of buffalo-meat, and a cheerful, busy little woman, who, to satisfy the gathering of visitors and her own little family, was cooking all the day long. We spent the time in the dug-out keeping the fire warm, except when paying a flying visit to the horses to give them hay or water, a business we accomplished with the best speed.

Three old bull buffaloes, "moss-backs," from the faded tint of their shaggy manes, were feeding quietly with the tame herd in front of the ranch. Sometimes one of the young steers would go up and butt a buffalo; this was all the attention the herd paid. The buffaloes heeded nothing; they fed during the day and left that night. An old bull is not worth killing. The hide is, from its size and the massive weight, difficult to strip. When off it is excessively heavy to carry, and it fetches a very much less price than the lighter and dark-haired pelts of younger animals. Although passing through the middle of the buffalo range, I never saw any great number; from sixty to eighty is the largest band I met. The hunters were all over the country, and the animals were terribly harassed. The slaughter is foolishly reckless; the hunters will kill them knowing they cannot, and do not, intend to skin them. The bodies are just left, and freeze hard—they are wasted. A hunter will ride in at dark and tell us at supper how, a few miles out, he met two cows: "I got off my horse and killed them," he was satisfied. The hunters are mostly men whose work is shut down during the winter. A few are professional—that is, follow no other trade.

The life of these hunters is ordinarily lazy, useless, and animal. If money admits of it, a whiskey-barrel has been brought out. They will hunt now and again on a fine day; the pleasure is to kill, the trouble of skinning is too great, and is avoided under any trivial excuse; every reason is good enough to keep them round the camp-fire. They grow slovenly in their habits, and dirty from carelessness. Their clothes are smeared with blood and grease; their hands and faces even grow strangers to water; the space round their temporary homes are filthy with scraps of meat, bones, and hide. Cooking under these circumstances falls into degradation. The vessels are not cleaned; each man marks his platter, etc., and uses it as he left it after the last meal. Great allowances must be made for the extremity of cold through which the hunters have to exist; and, being mostly men of but small means, in enjoying these outings they must be of frugal mind. A hunter occasionally has a thrilling adventure, in which his life is snatched out of a blizzard, or the icy grip of a north wind; often as not the peril into which he fell was due to some disregard of a well-known precaution through mere laziness. The romance of the life may be there still. I could not see it; but the idle freedom from all control, the needlessness of any regard for decencies of civilized life—these survive, and may be still indulged.

Though comfortable ourselves in the dug-out, our horses had a miserable time. The stable was unfinished; the logs had been laid one on another, but the interstices had not been filled; the wind blew through this, searching out the poor animals. At the end of three days the weather moderated, and we could all go on. There were some splendid display of Aurora Borealis during my trip, and once or twice, sleeping on the floor of a cabin, I was woke by the light which shone through the window, equalling that of early morning. The magnificence of the spectacle defies description. I remember more particularly two occasions: one, when the light was of a bright, rosy red; another, when from round a whole semi-circle of the horizon straight shafts of tremulous light shot upwards, and nearly met in the zenith; a long flame of white light flickering in the remnant of blue sky from the end of one of the fagots. There were a few herds of sheep near Miles City; these had lately arrived, and were being dipped to check the spread of scab. The sheep were launched into the shoot holding a solution of sulphur and tobacco in hot water; the temperature was 120°, and they were kept in four minutes. When they came out they were still panting with the heat as they stood on the dripping platform. Though the air was many degrees below zero the process did not seem to hurt the sheep. My trip was stretched out to the 20th December, when I thought it high time to leave travelling for another season. There was plenty of ice, but little skating. The children amused themselves, and the girls at times would induce one of the more obedient of their vassals to assist them, but the young men even of seventeen and eighteen looked upon skating as unmanly.

After the beginning of the year, I started to go along the line of the Northern Pacific Railway, which was open so far as the point at which it leaves the Yellowstone River, now called Livingstone, after having tried two other names. The wind coming through the canyon blows a hurricane, driving the snow till it lays the ground bare, then driving the sand and pebbles so that it is quite hard to face it. The hotel all night long shook and trembled. The inhabitants say this goes on the year round, which seems a pleasant outlook for settlers. The happiest suggestion possible was made by some one just before going to bed: "Suppose the house took fire!" One road to the Mammoth Hot

Springs and into the National Park left the railroad here; but since the completion of the line other more promising routes may have been opened. A sleigh carried us over the hills to Bozeman, a town prosperous already before the advent of the N. P. Railway. The latter was said to be at enmity with Bozeman; railways, as I have already said, acting the traditional stepmother. My wish was to go through into Washington Territory, which necessitated methods of travel changing between Concord coaches, open sleighs, and jerkies, that is of a wagon without springs. The Concord coach is the sanctioned design for a mail-coach. It is slung high on leather straps, has three seats, which neither can hold nor accommodate three people each; but nine persons are shoved inside. The windows are stoutly lined with canvas. It is a most ingenious torture-chamber for which you pay, and from which no one could escape in the case of an accident. The door is small; it is amusing to see the traveller, rolled up in clothes and buffalo-robes, forcing his way in and out. The sleighs were very simple affairs, open to all weather, and the pleasanter for being open so long as the weather was bright; but if snow threatened, the passengers generally scrambled to secure the back-seat, where they did not face the storm, and found a little protection behind the driver and his box; but in winter, unless very much pressed for time, the American will tell you you "don't want" to start out in bad weather. Falling snow is disagreeable enough; but the dangerous time is when a strong wind drives the already fallen snow in a thick cloud. It is then impossible to see a few yards ahead of you, and the chances are, whether mounted or on foot, you will lose yourself, and flounder into some drift. Every winter men lose their lives during these storms, which often precede a cold snap. There were two accidents of the sort on the road I was travelling, within a month after I had crossed over—driver and passengers all frozen to death.

The liveliest travelling was by jerky, the ordinary American farm-wagon without springs. You sat on a board laid across the wagon-box; that is, you tried to sit, for truly half the time you spent in the air, stiffening your arms to temper the bump, bound to meet your return to the seat. Whenever the snow was not deep enough for sleigh travelling the jerky was introduced; this was generally across some exposed high country. The road over the greater part of its length had only been—made implies some sort of artificial interference with nature; say—driven over since the railway had commenced its works; it was, therefore, splendidly rough, and of course frozen hard as iron. The driver would send his four horses along in great style, and we grinned and bore it. There was no weight to speak of in the wagon, and when going across the slope of a hill the wheels would slip downward, and the wagon would travel along sideways, the hind-wheels several feet further down the hill than where the team was trotting. At one place the only way to get down with safety, without unhitching and bother, was to make an S-curve on the face of the slope. It required clever driving to manage it properly; and, to speak well of the bridge that carries you over, I must give credit to the wagon for standing upon its wheels during the circus performance. It travelled both forward with the horses and slid very quickly down-hill of itself, the resultant being a diagonal sort of progress that made one wish to get out and walk. The steady bumping along the road is really hard carriage exercise; I don't know its equal in its line. Sleigh-driving, in a heavy machine, over the badly broken snow-roads, is not to be confounded with the sleighing in and about cities, of which we hear such delightful stories. On the road we travel there are many up and downs; we do not skim along the surface, but work over it with a continual crunching and screaming of the snow ground under the runners. "Sit heavy on the up-side!" warns the driver, when we all edge upward, those at the lower level putting their arms around the waists of their companions above; a nervous man jumps on to the edge of the box to give the up-side the full advantage of his weight. It's a sidling bit of road, with a steep angle to it, which abuts on a bank overlooking a stream, that only lives by the pace it dashes down the rocky bed. We are not afraid of drowning; but a turn-over into the torrent is not desirable with the thermometer below zero, and no habitations within a dozen miles.

The amount of clothes you wear when staging depends on your wardrobe. The best plan is to put on everything; this will lighten your portmanteau, and will not be one article too much on your body. You have of course purchased some special garments suited to the country, such as are never required in England; loose felt socks and soft overshoes for the feet; a leather waistcoat lined or wadded; the biggest fur-lined overcoat ever worn by man; you head and ears are well wrapped up, and a buffalo-robe is folded round your waist and over your legs. In these you feel as if you cannot move; but they will not always keep you warm. The feet somehow nearly always get cold. The bottom of the sleigh is hard and draughty; hay is sometimes spread, but at the first station where we may get out for half-an-hour, the cows regularly steal it all; it is no one's business to study the comfort of any one else. The driver is always a man of importance, and is a leader of opinion, in many matters quite outside horses and roads and his professional experience, at the two hotels which are the termini of his daily labors. "The stage!" is called out, and the hotel is in a bustle. It is the evening excitement; every one is interested, and the residents stare through the glass doors; the porter, clerk, and proprietor actually face the cold outside. Two or three bundles are assisted to rise and climb out of the sleigh; they are mostly hairy bundles, sprinkled with snow, on two props wrapped up in gunny-sacks. A small uncovered piece of face just shows eyes and a nose under which is a bristle of hoar-frosted moustache and icicles. The travellers are glad enough to come in, to throw off the robes and coats, and sit near the stove. After half-an-hour's time the driver drops in; he is substantially enveloped in folds of buffalo-coat, shawls, and flannel-lined garments. It is no joke to sit up and face the wind for eight or ten hours; and if twenty pounds of extra clothing are to help him to do it, he is bound to wear them. "A man is a fool, who can avoid it, to be either cold or hungry."

The road from Bozeman lay through Helena, which is the chief town of Montana. Here the Legislature was assembled, which kept the place lively. The two parties were equally divided ; and as the first business was to elect a president, clerk, lamplighter, housekeeper, and other officials, and as neither side would yield, there was a dead-lock. For ten days at least no business was transacted. The Legislature met and voted for a president; result—equal numbers for different nominees. After a short interval the members would vote again— same result ; after which they separated and adjourned, and so on day by day. Neither party would accept a compromise. The people looked on—first laughed, and then got angry, and readily accused the members of aiming only at their salaries. The number of sittings are limited, and the time of the members was consequently dreadfully wasted. After I left I understand some compromise was settled. The politics of a nation are always intricate to a foreigner, and the terms even are strange. A Democrat represented, conversationally, much what the term Conservative implies with us. Many Democrats do not admire a government by the people, nor are they violent partisans of republican principles ; they claim to include the more educated classes of America, yet their main supporters are the ignorant Irish of the cities. The elections this year have been going all awry, and promised a good fight for the choice of the next President. The American is always a politician so long as there is anything to gain. The vote is a substantial advantage which enables him to put in a friend, or to benefit himself ; the loss of citizenship is in consequence dreaded ; and a criminal, if he has interest, will be pardoned the last term of his imprisonment to enable him to retain this right. The elections which I saw went off much in the usual style, and apparently to general satisfaction, except of those who lost their bets ; but a few months afterwards I heard the whole lot of county officers were turned out by the government of the Territory at one swoop. To learn the worst one need only read the newspapers, which teem with accounts of frauds said to have been committed by the most prominent officials ; the openness with which such accusations are made, and the readiness to credit them, show that the people believe in the weaknesses of their own elected.

Although the citizen is very tolerant of the waste of county funds, he is constantly anxious that his particular State should in some way excel and appear prominently. He is not satisfied with having done well, but is thirsting to be appreciated and to extort praise. Nothing pleases more than the real or pretended astonishment of strangers. The barefaced flattery of the town and State by the traveller from the East, who does not care to provoke an argument with his interviewer, is printed in large type as sure to catch the local reader. Sensible Western tradesmen who have travelled in every State, and know New York probably better than the roads in their county, will remark that they like their Eastern friends to come among them, and be satisfied by their own eyes that they do not live in dug-outs or log-huts. As if the average New Englanders could be ignorant of the wealth and progress of great Western cities which they supply with train-loads of most inferior manufactured goods, and whose inhabitants never cease when abroad to sound the trumpet of defiance, and call all people who on earth do dwell to acknowledge the intelligence and enterprise of our citizens located in the special corner-lot from which they started. As for this assertiveness, one should admire it; it tends to the virtue of contentment. We English must plead guilty to a feeling of satisfaction with our corner-lot, and admit that we will constantly measure the results of creation, whether human or divine, by this standard. Once the youthful error of cosmopolitanism has been cast out, we may say nothing, but feel the superiority of our lot. But why should we alone remain silent, or be taxed with pride and prejudice, whereas the American, Frenchman, or man of any other nation receives from the stranger a patient hearing, generous admissions and when his fountain of eloquence has run dry, the kindly excuse that his patriotic vanity is to his credit? This is favoritism.

From Helena we went by Missoulah, after which we began to traverse a very wild country, and passed through a good deal of forest. The snow lay thick under the trees; along fallen limbs it makes a wall two feet high, and with flat faces coinciding with the width of the limb; on the stumps of trees a cylinder of snow is piled up. The scene is very picturesque, but mournful and dead. As we neared the terminus of the western end of the railway, which would take us to Portland, we heard startling stories of what was going on there. "Weeksville is very lively," the travellers said. Nine men had been shot or hung by the Vigilants during the past fortnight. Whatever the number might have been, the main facts were true enough. A little bit of a place, holding some twenty cabins or so, set down in the forest, had developed an amount of scoundreldom that was intolerable even to the tolerant men gathered at the end of a railway. A committee was formed, and notices were posted warning all people who were not following a trade, or were not gamblers, to quit. The exception in favor of gamblers was curious, unless you remember that out West gambling is a profession. The warning was signed with some mystic figures, thus 7—11—77. I don't know that I have the right figures; but none of the persons I asked could or would explain their meaning. The respectable people of Weeksville carried revolvers in their hands openly and in broad daylight. The two men in charge of the parcels express were armed with shortened smooth-bores, and kept most strangers out of the van in which the express parcels were carried; the alternative for passengers was to sit outside on a flat truck, exposed for a few hours to the night air. This was but on the link, for after travelling over some miles of new line, we were run into a station, and rejoined regular railway communication. We had at this stage entered Washington Territory. Our first experience of the train was to be blockaded in a snow-drift for over twenty-four hours. The snow was apparently as deep as up to the bottom of the windows in the cars. When we reached Spokan Falls we heard the line was breached in sixty or eighty places; a chinook or warm wind had produced a thaw, and the floods had washed out the line. After the delay of a whole week we reached Walla-Walla. Here we had come back to moderate civilization. There was a story that, in defiance of the law which compels all saloons to close during Sunday, the two principal saloons found it to their benefit to keep open on Sunday, and to pay the fine of $50 or so every Monday morning; it was rough on those whose business was not large enough to justify the expenditure. Walla-Walla used to be the headquarters of a great deal of stock business; but the country is turning its attention to farming. There is excellent wheat-land along the railway, of which the company had raised the price from $2.50 to $5, $7.50, and $10 an acre. There is plenty of timber and orchards. The apples, however, are in such quantities that much of the crop was not even picked. The price of cattle was very high; there had been a great demand for beef, and a two-year-old steer was said not to exist within reach of the railway. Cows, calves, and yearlings were offered at $20 a head all round. It was not worth while at that price to drive them on to the ranges I had left, where they were nearly as cheap; American stock-cattle being held at from $22 to $25, depending a good deal on the numbers of large or young beasts. A number of Canadians, and emigrants who had first tried the new Canadian provinces, were to be met throughout the northern part of the States; from what they said work was easier to find here than there. The winters are not so long, and the country being in a more advanced state of settlement life was not so hard nor so expensive. There is no doubt a good deal of impatience among the new settlers of Canada that their prosperity does not advance with strides equal to that of the American territories; and they perhaps consider the panacea for their trouble is a union with the States. Their real trouble seems to me a matter of climate. Where farmers may plough in June with greatcoats on, and find winter upon them in October, is not an attractive country; and until more southern-lying land is fully occupied the Canadians must wait. If they could transfer their farms into a more congenial climate, good; but the best of governments cannot lengthen their summers. Neither the French nor the new setlers are sentimentally loyal; they are looking to better themselves, and are ready to accept any measures. It is immaterial whether they call themselves by one name or another,—they are on the move.

Being about the middle of February the country was still covered with snow. There was nothing to be done in a small town; it was more prudent to spare one's endurance, and to await the spring in some better quarters. San Francisco seemed the natural place to do this. The easiest way was to train to Portland, and take the steamer down the coast, which was accordingly done. We encountered the most abominable knocking-about in crossing the bar of the Willamette River. The available depth of water here is only sixteen feet, so that very possibly the trade of this portion of the States will find a new outlet at Tacoma, or Seattle on Puget Sound.

Has anything been left unsaid of San Francisco? I think not. It is distinctly a pleasure-loving town, and a cheerful place for the stranger to sojourn in. There are plenty of theatres, and the bouquets and banners of flowers with which the impressible audience pay tribute to their stage-goddesses are the largest in the world. The Irish held a grand function in honor of their saint, and "processed" through the streets in disguises of Masons, Odd-Fellows, etc., bearing banners, and accompanied by much brass music. The cavalcade was headed by a "Leafy-yet" guard, the men and officer dressed completely as French soldiers, to the dismay of some wanderers of that nationality not quite broken in to American disregard of old-world *convenances*.

The travelling world has agreed that San Francisco is a charming place; but what are its charms, and how does it deserve the character? Other places have an equally good climate—at least, during the fortnight or month that at most travellers are able to devote to a single foreign city. We find mammoth and excellent hotels nowadays all over the world. Some of the streets, if fine, do not surpass those of other prominent cities. The private residences on the hill are from the outside picturesque, and by all accounts as comfortable within as wealth can make them. The wire tram-cars are exemplary in the way they carry you over the ground, smoothly and quickly; but there is no country within easy reach—the only drive is through the park to the Seal Rock. The harbor is a fine stretch of water, but nothing more. A few bare little islands unnecessarily block the channel without adorning it, while the Golden Gate, if an entrance was necessary, seems to have been put up in nature's most ordinary style. The first tendency of your traveller is to find fault; his imagination has been stimulated by the perusal of books made to amuse, or by descriptions of places from other visitors who, after having first given vent to their disappointment, revert to the second phase of the traveller's humor, which enables him to judge more correctly what he has seen, but arouses a new feeling—that is, to conceal from others to what extent he has failed in the commercial value of his voyage. It is a simple rule-of-three sum. If Switzerland and the Alps can be admired for twenty pounds, how much more must I seem to appreciate a country which has cost me two hundred? It is, I think, as a residence in America that San Francisco should stand high. A pleasant climate during twelve months is not to be found everywhere; while there is something about the social manners which is Southern—that is, expansive, light-hearted, and less strait-laced than those of Eastern towns. Speaking only from superficial knowledge, the difference between New York and San Francisco is not unlike that between London and Vienna as concerns out-door life and amusements. The dollar is king here as elsewhere, but there are some subjects slack in their allegiance, led astray no doubt by the gayer spirits of the foreign element, which comprises quite a number of Southern Europeans. Los Angeles is, however, running San Francisco a race for popularity; its size is increasing rapidly, and the romantic associations of orange-blossoms, laurels, and jessamine surrounding your cottage home form part of the stock advertisements of those interested in real estate in and around Los Angeles. The traveller and settler will do well to see both, and decide for himself as to relative merits.

California, after having been one of the best ranges for stock, is by degrees turning everywhere, except in the mountains, into an agricultural State. This of necessity follows from the greater profits of husbandry and the diminishing profits of cattle-farming to men of small capital. So soon as the soil becomes valuable, and the choicer portions are taken up by individuals, the cattle are no longer free to roam over the country, costing nothing for food; they must be looked after and herded; hay must be put up for their sustenance in winter, and a few days in the spring and autumn given up by the farmer and his boys are no longer sufficient for guarding his interests, nor for keeping track of his property, which are driven by the enclosing of their former pasture-ground to wander further afield. In the golden days of old, which in California are days of memory and not of tradition, the quantity of land actually purchased or taken up, whether under the laws or merely held by a sort of squatter right, would be limited to an occasional ranch along the fertile valleys of the big rivers, and to enclosures of meadows where the natural dampness of the soil or primitive irrigation gave large quantities of hay. The owners would let their horses and cattle run at perfect liberty to feed themselves, and only round them up when it was desirable to brand the young calves and colts, or to pick out horses or fat steers for the market. There are still a few wide ranges, the property of companies or of individual millionaires. The land is, however, owned, and if not fenced is constantly ridden over by the boys, who drive off outside cattle, and carry on a perpetual warfare with the Basque and Portuguese owners of bands of sheep which have to traverse the ranges on the way to the mountains or to the railroad. Those halcyon days of the California stock-raisers can never return. Land has grown exceedingly in value. Water taken out of the rivers is led by large canals over a wide tract of country; emigrants have crowded in, some purchasing small lots of twenty-five to forty acres at high prices from the pioneer farmers and far-seeing land speculators, who by ingenious manipulation of the land laws, backed by the power of ready money, have succeeded in acquiring considerable tracts at an earlier date.

So long as a State is but sparsely settled, the stock interest is sufficiently strong to make laws favoring that industry; but when the numbers of farmers have increased, the law-making, following the balance of votes, is taken into the new hands, and one of their first acts is naturally in the direction of safe-guarding their pockets. Whereas before the land-owner had to protect his crop from the roaming herds, subsequently the stock-raiser is held responsible for any damage caused by his cattle, and therefore has to look to this. Practically it is found convenient by the farmers to protect themselves, and, either in combination or singly, they soon begin to enclose the land where the more valuable crops are grown, and in the older settled districts fencing is the order of the day. The cattle are thus shut out of the water, and lose the protection of the copses and fringes of trees which border the valley streams. They leave the bottoms and range far back in the mountains, where they find small springs, and put up with the shelter of broken ground. Formerly timber was cheap, and it was mostly used for fencing, but now barbed wire of different patterns is more common. The wire generally consists of two strands loosely twisted, with small knots of wire in pairs at short intervals; each knot is given one turn round a wire, the ends project half an inch, and are cut off diagonally so as to leave sharp points; thus at about every foot there are four barbs; it is impossible for any animal to squeeze its way between the strands without being torn in the process. The laws of the States provide that the wires should be at some fixed intervals, and at certain heights from the ground; that the standards which carry them should not be more than a certain distance apart, and that a top rail or plank should be fixed so as to warn stock of the presence of the obstacle. These rules are often not closely observed, and it is a common sight to see valuable animals badly torn by having unwittingly at night come in contact with barbed wire, or by having struggled to make their way through an easy place to reach water or better feed.

The laws which concern stock, though they differ in the various States and Territories, have been in each case made by people who know exactly what they want; from the local standpoint they are excellent—that is, they suit the majority and benefit the framers. This, no doubt, appears the best ends of justice to men struggling for wealth in a primitive society; the basis of equity may be neglected, each must look after his own interests, and if a man does not like the laws he can move off. If stock-owners are in power, they say to the small rancher, "Fence your fields"; if the farmers are numerous, they turn on the stockman and say, "Herd your cattle," while all combine against the stranger within their bounds. Laws are useful to those who command the market, and can thereby profit themselves or frustrate the commercial competition of outsiders; at least, such is the hearsay evidence of the inhabitants, and one of the leading topics of their newspapers. It is a common saying that the rich man may secure a verdict. With all this fencing and irrigation, the lawyers in California have their hands full of work, and a harvest which lasts all the year round.

The ranchers living further down a river find the volume of water on which their crops and stock depend gradually diminishing as the upper reaches are settled and new canals are laid out. Suddenly, in some particularly dry year, there is no water at all in the lower channel of the river, the crops suffer, and the cattle must be driven to the hills. In the old days the injured party was apt to start with his shot-gun and argue the matter in person; now the majesty of the law favors the long purse, and the man who wins his case recovers just enough to pay his lawyer. There is no more fruitful source of litigation than water rights, and in purchasing land the buyer must be extremely careful to know that his title to water and to a fixed quantity thereof is undoubted, otherwise he may be called upon by his neighbors to join in a lawsuit to protect their common rights, or perfind that he has bought the privilege to fight single-handed a large owner who has strong influence in the courts, and is prepared to appeal as a pure matter of business.

A great many tracts have lately been laid out in plots which, being contiguous, and in a way connected, are termed settlements. These plots have been a good deal bought up by continentals acquainted with vine and fruit cultivation. Possessed of very little capital, and accustomed to labor with their own hands, they are industrious and thrifty, and are the best present emigrants into California. Owning only small parcels of land, which are highly cultivated, they are building up the future prosperity of the State, which will hereafter depend less on wheat and more on wine. There are many large vineyards cultivated by Chinese coolies which are extremely well laid out and carefully tended. These must have cost great sums of money, as the land has to be cleaned and levelled. After the vines are planted they must be watered and the soil kept clear of weeds; in three years' time you may expect a crop; it is not till after five years that the garden is in full bearing. The reports of the success of vineyards are not to be accepted too literally; the fashion has been lately to cry up viticulture as a certain fortune, and fabulous accounts are told of the present wealth of certain individuals realized from small beginnings. A great deal of this must be discounted as the patriotic exaggeration of Californians; the returns hinge of course on the outlay of the first few years; of this the chief burden is the original cost of the land. Those who bought at an early period of the settlement of the State, or some years back, when a monetary crisis brought down the price of improved estates, benefit *pro tanto*. The new settler must be prepared for a large difference in percentage between that earned on estates for which at the outside $2.50 an acre have been paid and what he may expect to receive after laying out $25 to $40 an acre.

Many are and will be led away to settle in California and go in for fruit-farming or vine-growing by the glittering stories which appear not only in newspaper paragraphs, but are also solemnly detailed in official reports with all the support of elaborate statistics. The cream in both cases has been skimmed; there are already a great number who have started in the business. The long delay before any return can be hoped for is a great deterrent to men who must live by their labor, and whose principal capital is their hands; the work requires some experience and unremitting attention. The valleys where the fruit thrive best are often exceedingly unhealthy; both in the Sacramento and San Joachin valleys there is a local fever accompanied by ague, which saps the strength and energy of a man;

it may be fought against for a few seasons, but in the end necessitates a complete absence for recovery. Whether due to irrigation, or to the chills of evening which follow on days when the thermometer ranges high in the sun, the fever is there. The vine-grower's work is one of exposure; there is unfortunately no such process as acclimatization; the old settler of course avoids preventible risks and the foolish mistake of the inexperienced, but his maxims as a rule have been earned at a costly price of health. It is no uncommon occurrence to hear of the farmer, his wife, and children being down with the fever together. A home at the foot of the hills or near the coast may not be surrounded by soil as fertile as that in the great valleys, but the fruit-trees thrive, and the vines are said to produce on the mountain-sides a better wine-grape; the corn and oats are superior in quality, if not so abundant in bushels per acre; the climate is most enjoyable and healthy, while the scenery is charming, having the merit in English eyes of looking somewhat like England. In California trees are seen growing scattered over the country, while east of the Sierras you might say there are no trees on the plains except where they fringe a stream.

When cultivation sent up the price of land, owners could no longer afford to retain arable land as pasture; the alternative of sending cattle and sheep into the hills for summer grazing has its risks, so that small owners of stock cattle have nearly disappeared, and most farmers have for several years past been reducing their flocks of sheep.

Among the diverse methods of making and losing money peculiar to America and the far West, one of the more speculative, at the same time interesting from its slightly adventurous and strongly nomad existence, is that of driving stock. The profit lies in that golden rule commercial, buy in a cheap market, sell in a dear ; but between the two transactions there is a wide field to traverse. During this interim money goes out at a great rate, and the stock run various risks which tend to diminish their numbers and condition. Your best efforts are therefore made to watch the herds, to prevent strays, to see that they get sufficient food and water, and not to over-drive them ; and when you succeed in selling your herd, the price obtained may or may not requite you for all your trouble. Texas has lately been a good outlet for some of the surplus stock of California ; young sheep have been bought and sent by rail half-way, and afterwards driven into that State. For many years previously large bands have left both the northern and southern parts of California for the newly settled territories of Colorado, Wyoming, and Montana. The numbers run up to many hundreds of thousands each year. The bands start from every county, but generally cross the Sierra Nevada over three main passes. The pass north of the Central Pacific Railway is the outlet for sheep from the Sacramento Valley ; southeast of San Francisco the sheep cross a little north of the Yosemite, while those from the direction of Los Angeles turn the lower end of the range, and, taking a northward direction, subsequently join the second route. This second trail joins the first near the head-waters of the Humboldt River ; from here the trail crosses a corner of Idaho and Utah, and splits ; one road leads north into the western portion of Montana, the other goes east into Wyoming and Colorado.

The months of January, February, and part of March, 1883, had been very hot and dry; the absence of usual rain threatened the farmers with a drought, of which one incident would be the scarcity of grazing, compelling all who owned stock, whether cattle or sheep, to drive their herds into the mountains, or to sell. Either of these alternatives is a matter which admits of little delay. If rain does not fall, the sparse grazing to be picked up in ordinary years along the road, on which animals must depend while travelling, has totally disappeared after the passage of a few herds. There is nought but dust, under which sheep for a time will continue to find scraps and pickings, though not a blade is observable to the eye; this of course does not last long. To buy sheep in such a season is a mere lottery; rain may fall, when your transaction turns up trumps; rain may hold off, when your sheep, unless singularly well managed, will weaken, and, once they begin dying, depart by hundreds. Fortunately for the country, the rains, though tardy, fell at last; prices rose and fell to the content of all men except the speculators of San Francisco, who lost their stakes on the prospect of a bad season. Though bad for the country, the weather in town was all that it should be during a man's holiday.

After inquiry in one or two places along the South Californian Railway, I thought I might invest in a herd of sheep and drive them over into the territories. Fresno seemed as good a place as another, and making this town my head-quarters, several days were spent in scouring the country and inspecting any band of sheep that was for sale and likely to answer the description wanted. The better-bred sheep have been mostly improved with Spanish merinos; they are small-sized sheep, but carry a heavy fleece; they are thought more hardy than French merinos, and are close feeders, finding something to eat on the most barren-looking plains. Although well in the middle of April, the weather, owing to the late rains, was not settled, and while looking at a flock which had been taken up into the hills we were caught in our summer clothes in a storm which blustered and snowed from noon till seven o'clock, drenching us to the skin. We reached our little lodging in the village about eight o'clock; supper was over; it was not to be expected that anything afterwards would be cooked, so we found what comfort we could in cold remnants, and blessed the independent spirit which scorns to consider others, and makes him who wants the abject slave of him or her who has. Money is often not a power in this land of monopolies. There is but one meal cooked; if you are hungry, pay and sit down; whether you are a millionaire or a wood-chopper, the price is half-a-dollar; but you must want what the good lady has got ready, or you can go without. I was presented with my bill and a request to leave a fair hotel in a good-sized town, where I had lodged several days, because, being late, yet within the fixed hours, I could not eat a lukewarm supper, but went over the way to a restaurant, where I got a couple of chops and an omelet. I had been dissatisfied! was the verdict of the waiter.

Finally, a couple of bands of sheep, numbering about 5000, were bought and paid for. Two certificates that taxes for the year had been recovered on them were obtained from the county office. These were the most informal documents, merely stating that Mr. So-and-so had paid his taxes that year; nothing was added to say that the sheep were those now my property, that they had any particular mark, and one certificate was not dated. I will, however, speak well of them, for I was once called on to show my tax receipts, and after some very proper objections to the informality of the documents, they were allowed to pass. People moving from one neighborhood to another should carry their receipts along with them, as they are liable to be stopped wherever there is a collector, and show cause why they should not pay the county taxes on the value of horses, wagon, and outfit, and something in the shape of poll-tax on each individual for roads.

It is difficult to know how you stand in this strange Western country, which combines the elements of the wilderness with the civilized refinements of elaborate laws and taxation. You admit it certainly is wonderful, while you ask yourself, Where does the money go? You see schools almost everywhere; this rises to a hobby with Americans. State education saves their pockets, and in each new settlement the people are anxious to increase the number of families to the quota which will enable them to start a school and share in the expenditure of the county revenues. It was very pleasant to see the little folk riding to school, sometimes two on a horse, or walking with their books and little basket. I asked myself often how the "school ma'am," young enough herself, kept discipline among and conveyed instruction to the mixed gathering of children, boys and girls of all ages, up to sixteen and seventeen, in the single room of a small village school. That severe measures are applied we can guess; that the parents approve necessary severity is no more than to be expected from people of sound common-sense. I heard at one place sincere regrets at the loss of a master who had advanced to better things; he was a great disciplinarian, and had commenced his taking charge of an obstreperous lot, that had run wild under less firm hands, by whipping the big girls. The story was not told me, but was introduced in a little gossip over school affairs among half-a-dozen men smoking in the veranda; it seemed to me so improbable that I joined in, and satisfied myself that the story was an accepted fact. The master certainly deserved all the credit for his courage; one would have liked to see him do it, for American girls are not meek-spirited; conscience, and a feeling of getting no more than their deserts, must have overpowered that lot.

Besides the sheep, it was necessary to get an outfit, which consisted of a wagon and pair of horses, two riding ponies, cooking and eating utensils, saddles, harness, a few tools and a stock of food to start with. When the boys shall have thrown their bedding and bags in the wagon, the whole will make a solid load for the team. The wagons all over the West are imported; they are very much alike, whoever are the makers, and vary mainly in diameter of wheels and size of axle. The driving-seat has a pair of springs, and hooks on to the sides of the wagon-box; the body is painted green, the wheels and working-parts red. You will see them in dozens at most railway stations, lying in parts; they are quickly put together, and there is a large demand for them. They are much lighter than the ordinary English farm-wagon; but then they are weak, and do not last, which is due to the hastily-dried wood of which they are made. The usage they receive is rough; they are frequently loaded far beyond the maximum which even the makers will guarantee, and rattled along with four horses by a reckless young fellow, caring neither for his master's property nor his own neck, over a nominal road with ruts and wash-outs and boulders. But our lad has driven from the time he could hold the reins; he is at home on the box; perched up there, with one foot dangling over the side and resting on the handle of the brake, he sends the team along. The wagon leaps and swings and sidles, steered as well as may be past the big boulders, and checked through the wash-outs by a heavy pressure on the brake. The journey is lively, and the driver has quite a time in recovering his seat when thrown out by a jolt, or slid to the further end by the sway in turning a corner or changing his ruts. This is something like driving, and, as a science, far ahead of any skill called into play in the jog-trot travel along our humdrum and excellent roads. When arrived at destination and unloaded, the wagon is left standing exposed to the weather in front of the empty shed.

The method of harnessing is quite different from the English plan; it has the recommendation of simplicity and saving of trouble, which latter virtue, if not the mother of this invention, has been a kind patron. You first pass the collar, which opens above, round the neck, and buckle it; this is more practical than pushing the collar over the head, which would be impossible with a backing horse in the open, and terrifying in the numberless cases of half-broken animals. The harness—saddle, traces, and crupper, or breeching—are then thrown on in one piece, and the hames are strapped at the bottom of the collar. The head-gear is then slipped on, the inner reins crossed and buckled to the bits; the pole, or tongue as it is usually called, is lifted, and the breast-strap is buckled, which suspends the cross-bar to the collar; the belly-band is strapped very loosely, and the traces made fast. The process is slow in describing, but very expeditious in fact. In unhitching, first undo the traces and belly-band, unbuckle the breast-strap and inner reins, take off the headstall, which hang on the hames, and make fast to the ring by a turn with the rein; unbuckle the hame-strap, and sweep the whole lot on to the ground in a heap, on which place the collar; the horse is free. The harness is very seldom cleaned; it may be greased once a year.

I brought a tent along, but it was not pitched more than twice during the whole journey. A large sheet of canvas, which served as a tilt to the wagon in rainy weather, was eminently serviceable. On the plains, where nothing stands higher than a bush, which hardly gives shade to a dog from the hot sun, this canvas was stretched from the wagon-bows to pegs in the ground, and gave us a little shelter while we ate our dinner and wearied in the long sultry hours. A mess-box was fitted into the hind-end of the wagon; it was fitted with shelves, and held a supply of daily wants; the door hinged at the bottom, and when lowered was propped by a stick, and made an excellent table, on which food could be prepared for cooking, out of the dust. But we ate our meals on the ground, as there was more room for everybody; besides at noon we wanted the shade of the tilt, morning and night the light and solace of the camp-fire.

The important affair still left undone is to hire men. Settlers in California have come to employ Chinese labor almost exclusively for indoor work, and to a great extent for any outdoor work which is continuous; not, as one might suppose, that there is an economy therein,—I should almost think the contrary. The Chinaman is a thoroughly self-satisfied being; he considers his work "all same as 'Melican," and lets you know that he is not to be hired for less than white man's wages. With due regard to the present spirit of tolerance which checks any blame, lest we err from insufficient knowledge, in canvassing foreign nations, I would assert that Chinese labor is neither in quantity nor quality equal to that of the average European. All over the world the Chinaman is a copyist; he invents nothing and improves nothing; his aim is to produce a fac-simile, he can never excel. Notwithstanding this inferiority, he is preferred because he is more to be depended on, mainly in the matter of sobriety. As a household servant he looks clean, is fairly willing, but far behind the class of domestic in European houses on the other side of the Pacific. The American atmosphere of independence is certainly inimical to good service; it has breathed into Johnny the spirit of equality, and makes him careless and bumptious; nevertheless he has a solid footing in California, and you find a smutty, yellow-faced cook in small farm-houses, where elsewhere in the States the wife and daughters do the household work. For what good purpose this assistance sets the women free is not easy to guess; rocking the chairs seems the most arduous duty in many Californian homes, and it is one which is faithfully carried out.

For riding, driving, and the heavy work during a few weeks' harvesting, Americans cannot be beaten. Self-reliance is their prominent characteristic; every man will undertake any class of work without any previous training, and with the greatest self-assurance will proceed to lose your sheep, smash your machinery, or spoil your crop, for thirty dollars wages and all found. "You cannot teach him anything," is a common saying among the boys, which hits off very well the exaggeration of a great good point; for to this self-reliance is due a great deal of the wonderful advances in America. It induces the men to widen their experiences, to turn their hands to all trades, and to start off on the longest journeys, trusting in themselves to pull through. One result could hardly be avoided—that is, the number of indifferent workmen all over the West. A boy of eighteen attaches himself, say to a blacksmith, wheelwright, and wagon repairer; this trade would in most countries require at least a couple of years' apprenticeship. Not so with our intelligent citizen; after six months he will try to boss the shop, or start an independent concern in a neighboring town.

This would not answer in the East, but forasmuch as the numbers of people settling out West are continually demanding a proportionate increase of the trades, the young fellow will probably hire an industrous German or Swede who does know the business, and the chances are the trade thrives. If he is now steady he is nearly bound to get on. After a few years he marries, runs for the county constable, lends his surplus money out at 24 per cent, and will have in the mean time located a ranch and bought cattle. Before he is forty years old he has sold his business, has been for some time living on his farm, and is worth—well, it is hard to say—there is one sum on which he is assessed for taxes, another and considerably higher one which sounds better in conversation.

Unfortunately whiskey bars the road to prosperity. The career of the steady and lucky apprentice, though not unfrequent, is not the rule. In a country where there are no amusements, sociableness is found at the saloon; whiskey and gambling are the only possible excitement. A steady man sets up for himself in business; among the remnant you must select your help. An American servant will work satisfactorily for a month or two, and has some little money in hand; he has been absolutely abstemious all this time. Suddenly one day he is not at his work, and, to your great inconvenience and annoyance, you hear of him on a spree in the nearest town, or perhaps in the lock-up. By-and-by he returns. He is neither bold-faced nor penitent; what he has done concerns himself, and is a matter not for other people's criticism. He knows he may have to go; that is your right, which he would not for a moment discuss. He receives the balance due to him, and invites you to take a drink. There is, indeed, a certain dignity in his behavior, if one can apply the word to so small a transaction.

In some cases the best-intentioned man, somewhat to his regret, if he has any objective feeling in the matter, is liable to place his employer's property in a risk which no temptation but that of whiskey would induce him to allow. He may have been sent out in charge of a herd of sheep, a work he understands and does not dislike; it is a solitary but not an arduous life; there is plenty of grass, and no great distance to travel to and from his hut; he has a couple of good dogs. His working-hours are spent sitting or sleeping in the sun or shade; he has leisure to cook, and, if he pleases, to read. The foreman comes round once a week to replenish supplies, and reports that the sheep are well looked after; you feel at peace, and, in this happy frame of mind, resolve that to preserve this treasure you will increase his wages, possibly give him an interest in the herd of which he has taken charge. But the day of reckoning comes; a passer-by informs you that Bill or Jack has been drunk the last two days and treating all comers at some roadside saloon twenty miles from his range. Your first feeling is concern for your property, but nearly simultaneously you experience an emotion of annoyance that an otherwise good man, who might get on, is so weak before the blandishments of bad whiskey. You mount and ride out, to find matters as they have been told you. Bill, if not asleep, is in great good humor. "The sheep are all right—all right—somewhere on the range; come and have a drink." In the mean while two or three thousand head, with no one to look after them, are liable to wander into a neighbor's young crops, or to scatter themselves in bunches over the hill-tops. You are fortunate if you can spare a man in this emergency. Maybe Bill has had his drink; he is not mean, and will offer to return and collect the band while you ride to town for his substitute.

In America there is neither giving nor receiving even so much as a week's notice. If you offend your man he asks you to settle, as he guesses he will quit; on the other hand, if he annoys you, you are at liberty to be rid of him at once. The rule is a good one; the circumstances of the West make it no hardship on the working man; the market is still very much in his hands. The employer occasionally has a rough time. As a class these may appear to their hands exacting, principally those who have come from among the laborers themselves. Such a man will rise an hour earlier if he has hired help, but even he will, I fear, fail in getting his money's worth.

No wonder, then, that so many employ Chinese; the latter, having been taught to work, if they like their situation can be relied on. You need fear no jinks, but the work must be more watched. They have no great invention, their labor is mechanical and routine, an emergency finds them unprepared; they will also be careless and dirty, and even sometimes maltreat the stock. The balance of good and evil is, however, in their favor; men who have employed Chinese successfully quickly make up their minds to use them exclusively. It does not answer to try to work with both races; the two colors will not mix pleasantly, though it does happen, as for instance in a woollen factory in San Francisco white men and girls were working together with Chinese at different looms in the same room. If Americans are driven out of the labor market, it is owing to their own intemperate freaks.

The general sample of laborer in California is lowered

by the tendency of all vagrant and idle good-for-nothings to go West. The loafer's paradise of four dollars a day with little to do recedes towards the setting sun. Tired of steady work, he packs his traps, and starts for some large town in the next State. Impelled by the spur of want, he takes an odd job, and for some time lives in a hand-to-mouth style. This does not fulfil his dream, hearsay tells of better things out West. He sells his kit and blankets, and reaches some mining district where good wages are going. Here the work is too hard to please, so, having saved enough to pay for a railway ticket, off he goes again. Another trial, and another disappointment. He is now afoot, and, with some similar companion, trudges long marches, eking out his few dollars by small jobs, suffering actually hardship and semi-starvation, and working half as much again in getting over the ground as would earn him a good living at any of his halting-places. At last he reaches California, which still retains some of the nimbus which shone around it in the glorious days which followed '49. Times have really altered; he finds himself facing the melancholy fact that wages are not so high here as they were several hundred miles back, that the golden sun sets in the ocean still further West.

Among such men and those who bid for their votes, or who find in them customers for their goods, their whiskey, or newspapers, there is naturally a great outcry. The Chinaman is underbidding the white man; he helps the monopolists to grind the faces of the poor; he keeps down wages; he impoverishes the country by taking his earnings out of it; he is not a citizen,—added to a string of stock complaints which a good memory and the pleasure of hearing himself speak enables most Americans to declaim. A law has been passed to limit further Chinese emigrants to those who are traders, or had previously been domiciled in America; but the "heathen Chinee" is too subtle; they come to San Francisco provided by the Chinese authorities with traders' certificates that are manifestly irregular, while numbers are said to slip over the border from the Canadian provinces. Any one residing in California will readily sympathize with the people, who find their beautiful State overrun by this alien race, whose customs will not assimilate, whose uncouth appearance and habits on a lower level, joined to arrogance and impertinence where they can display them, naturally awaken prejudice This feeling is of course wrong viewed from the heights of philanthropy, so nothing can be easier than in a New England home to dilate on the injustice to the Chinese, and the reversal of America's grand title to be a land of freedom to all. If you cannot throw off this ill-feeling, but can get rid of the object of your prejudice, it is human to settle the matter comfortably to suit yourself. The absence of Chinese coolies, and consequent loss of labor, would certainly be a detriment to the State in delaying its present rate of expansion; but the Americans have gone fast enough,—it would be no harm to hold their breath and take stock, and see what sort of country, with what real development and debts, they mean to hand down to their children. The desire to exclude the Chinese race stands quite alone; there are many resident strangers, French, English, and German, in the higher walks of commerce, whose purpose is to make a pile and withdraw; and among the laboring classes.

Portuguese, Basques, Italians, and others practise the same habits as those wherein the Chinese offend—namely, they live frugally and save their money, spending as little as possible in the country, and eventually taking out so much as they can.

The American does not like foreigners, but he tolerates their presence if they will follow his example and adopt his institutions; but to be a separatist, to live in small national colonies, to appear or behave differently to the accredited type, not to care for local topics or the politics of the saloon—these are all crimes which the American cannot allow. You are welcomed to the country, but you must "fuse" and learn to think and act as a good American. I would illustrate this by referring to the local feelings awakened by the Mormons in the territories through which they have spread, and the general attitude of Americans towards negroes all through the West. The former are men of the same race, and precisely alike in every way to the mass of Western settlers, but from whom they hold themselves aloof and quite distinct under an idea of spiritual selection; they are heartily abused, and every effort is made to injure them. The negroes, who compete with the white man in every class of trade and work, now number nearly one eighth of the whole population, and if emigration from Europe slackened they would increase more rapidly than the whites. They are an alien race, but they are adaptive, and copy the Americans in their ideas, religion, dress, food, language, and prejudices. There is no strong antipathy to the negro which would wish to send him back to his native Africa, though I doubt not but this will follow.

It seems impossible for people on the spot to be ruled by abstract ideas of equity. If fair play to the stranger means a hardship to your neighbor, and may prove a cruel disadvantage to your children, are you then to do right and let the skies fall? Which is right? If by an inevitable law the weaker perish and the strong survive, why give points to the adversary, and, in carrying out some theoretical doctrine of right, make the battle more difficult for men of your kind? Some such arguments may apply to this race antagonism in California. The Americans are a practical people, and go straight where their interests are concerned. They will undoubtedly settle the question to their own ultimate satisfaction, and to the advantage of the Anglo-Saxon races.

In choosing sheep-herders, the best will be found among the Mexicans, Basques, or Portuguese. These two latter do not, as a rule, take service except with their own people; their aim is ultimately to possess a share in the herds, and to rise to the position of owners. The Mexicans enter into service willingly enough, but dislike to leave the temperate parts of California. It is a great advantage when employing them to be able to talk Spanish. They can seldom be persuaded to join a drive which takes them off into unknown regions; they are profoundly ignorant of the geography of the world beyond their districts. There is, besides, little inducement to travel with stock for good men, who are sure of employment locally; they have to undergo hard work, exposure, and some privation; and for what result? None! Every cent a man can earn above ordinary Californian wages will go to pay his railway fare, even by emigrant train, on his return to California. The rates over the Central Pacific Railway are excessive, and a severe tax on commerce and industry.

The herd required about six men besides a cook, an important member of the outfit. Rather a scratch lot were got together, mostly men whose purses had run dry, or who wished to leave the district for private reasons. It was impossible to find any one fitted for the post of foreman; not from backwardness of persons ready to take the part who are satisfied that they will "fill the bill"; but if you exact a fair amount of previous experience, and some knowledge of the road, it is hardly worth while trying. Besides, the men never get on well with a foreman; each one is pleased to think that he could "boss the job" much better, and will proceed to try for his own way. It is almost better in every case to get the best men available and manage the business yourself. You will be sure of your own interest then being considered first, which would not be the case otherwise.

Some twelve or thirteen hundred sheep had not been clipped that spring (in California the sheep are shorn twice a year); it was necessary to take their wool off before starting. The band was driven out on to a barren plain, where a few tumble-down open sheds guided you to the shearing corral. The first thing to do was to go round and rearrange panels, make fast ties and block holes, so as to keep the sheep in the pens. A mixed band of Mexicans and Chinese did the shearing, each man careful not to catch any sheep which on account of size or wool was likely to prove slightly more troublesome. A badly boarded floor was all the men worked upon; the fleeces, having been rolled up and tied, were thrown outside. A strong wind, bringing clouds of dust, was blowing the whole time, and scattered tufts of wool all over the place; at least five per cent of the wool must be lost in this careless, haphazard style of neglecting appliances and saving in first outlay. The fleeces were thrown into a long bag hung on a stand, and were filled in by stamping on them; the bags are then carried to the railway, either sold to brokers or shipped to an agent in San Francisco.

About the 23d April preliminaries were completed, and the herd started on the road, which lay at first along the railway running through the San Joachin valley. The sheep each morning were divided into two bands, and kept about half a mile apart. As the land is all owned, the drover has no right beyond the width of sixty feet. Where there are no fences it is futile to attempt to keep a large flock within such narrow limits. The sheep spread across some two hundred yards, and so long as they are kept going it is hoped that the landholders, most of whom are owners of sheep which have to be travelled twice a year, will not object. As a rule the large owners used not to trouble travelling bands much; but a small man, whose land borders the road, mounts his horse on the first sight of the column of dust which announces the approach of a band of sheep and rides to meet it. He is all on the fight; first he wants you to go back, then to go round, and last to manage the herd as you might a battalion of soldiers, and march them past his grazing ground in a solid pack, on a narrow strip of road. It was a lucky day's travel in which you had not to go through some annoyance and jaw.

The land-owners are quite right in objecting; a band

of sheep passing by does no good, that is certain; whether they do harm I do not know, unless a very large number of bands are travelling. In places where fencing has been put up, leaving the regulation width of road, there is a good deal of grazing to be picked up in a day's journey ; the residents, however, generally try to clear this off early in the year, and turn their stock on to the road to save their pastures. Each year driving becomes more difficult, grazing increases in value, the fields are fenced ; land is also more broken up. It would be difficult to take sheep on the drive, close along green crops, without their breaking into them. Here troubles begin with the farmer's opportunity of claiming compensation. As a matter in which he may have to go to law, he must exaggerate the damage. He will always find neighborly friends who will swear to his complaint, and assess the loss arising from a few hundred sheep crossing a corner of his field at the price of a crop from twenty acres of wheat. Of course to the drover it would be a great trouble to come back a month hence, bring his witnesses, and fight the case in a hostile court ; his delays may be endless, as Americans are past-masters of chicane ; he is therefore bound to compromise the matter to the least disadvantage he can.

Before taking the sheep out of the country it was necessary to dip them to check scab ; the Californians are not over-careful in eradicating this disease. I did not hear of any practical system, as in Australia, for dealing with the malady, or for detecting its presence in certain flocks, and compelling the owners to effect a cure. Most owners dip their sheep at least once a year, after shearing, but hardly in any band you pass can you omit noticing marks of the disease on some of the sheep. In some of the territories laws have been passed, and scab inspectors appointed. The attention of the latter is directed mainly to overhauling bands passing through; provision is generally made by the county or State to pay these individuals. I only met one on the journey, who asked for certain fees, on which afterwards he did not insist, owing, I thought, to the presence in camp of another man belonging to that part of the country. I may have been wrong, but I was suspicious at the time.

The use of a dipping-station was offered me: to reach it the band was turned off the main road and driven towards the foot-hills. The principal part of such buildings is a trough lined with wood, twenty to thirty feet long, five feet deep, and two-and-a-half or three feet wide at the top. This is sunk in the ground. At one end is a shed roofed over to shelter the men at work; the floor is boarded, and has a slight slope towards the trough. At the other end the sheep walk out of the trough by an inclined plank on to the dripping platform, which is divided into pens. This is also boarded, so that the water which runs out of the fleece may fall back into the trough, and save material. At either end are enclosures to hold the sheep which are being worked; iron tanks for heating water stand conveniently near, as with some of the scab-curing ingredients hot water must be used. The number of the sheep which can be handled in a morning are folded in a large inclosure; smaller bunches are cut off and penned up near the shed, which itself will hold some thirty or forty sheep. So many are driven in as to crowd the place tightly; the gate is shut, and two men step in, standing near the outlet which overhangs the trough. The sheep naturally turn their heads away, and press more closely to the upper side. This is just what is wanted. The men catch them one by one by the hind-leg, with a good pull and final jerk drag each one towards the trough, turn him round, and tumble him head first into the fluid. It is rough work, but gets through the business at a fair pace, which is always good enough out West.

When properly done the sheep souses into the trough head first, and comes up turned in the proper direction. Seeing the others swimming in front, he follows, and walks up the sloping plank on to the dripping platform. Sometimes it happens that a sheep will fall in backwards, and floats feet up in the air, feeling no doubt particularly bad with the composition of the dip, half chemical, half turbid with grease and mud out of the fleeces, filling his mouth and nostrils. A man stands alongside the trough armed with a long pole with a crutch at one end; it is his duty to restore these acrobats right side up, to push the heads of those not properly wetted under water, and to keep the line of bathers moving on. When one compartment of the dripping platform is full, a gate is shut, and while the alternate pen is filling, the former lot of sheep stand and shake themselves, sneeze, cough, and generally strive to recover their mental equilibrium. Soon their turn arrives to be let out into the larger enclosure. Here they ought to remain till nearly dry, as the dipping mixtures are more or less poisonous, and should not be scattered on the feeding-ground, as would happen from still wet fleeces.

The dip mostly used in lime and sulphur, which is effective in killing scab, but makes the wool brittle; it has the merit of cheapness. A decoction of tobacco and sulphur is also common. Both of these have to be applied with hot water, which is a great additional trouble, as the appliances at most dipping-stations are of the rudest. A weak solution of carbolic acid and a patent Australian chemical are also used for dipping; these can be mixed in cold water. Some men put their sheep through the natural hot mineral waters which abound in the West. Each farmer will swear by his own particular spring. It cures the scab in sheep, removes corns and rheumatism in men, and is efficacious universally; he nurses a pleasant dream that some day its virtues will be apparent to an Eastern capitalist with money to develop it, and to create an establishment like the White Sulphur Springs, with a vista of shares, purchase-money and a snug monopoly for the rest of his days.

There is nothing the Western man calls out against more than he does against monopolies, yet five minutes after a tirade, in which excellent principles have been laid down, and the injury to the country and its citizens dwelt upon from railways, holders of large land tracts, owners of bonanza mines, and millionaires generally, you find that this ardent Republican has either a quartz ledge or a water privilege which he is waiting for some Eastern capitalist to develop, or that he may be an arrant monopolist himself in a small way, having secured from the county the right to levy tolls on a road which has not cost him twenty days' labor, or he has fenced in some natural curiosity, or has enclosed the only water to be met along twenty miles of road travelled years before he settled in the district. In truth, everywhere the public good is sacrificed to the interest of individuals.

About twelve o'clock the sheep penned in the morning are through; the men knock off for dinner. Although there are three reliefs in plunging the sheep into the dip, it has been hard work. The sun is very bright and hot, the air is close inside the shed. The work of driving the sheep into compact bunches in the pens is tedious, and when you have jerked forty or fifty sheep by the hind-leg you find yourself winded, and your back aching. Our only interruption was an inroad from a coachful of tourists, who had just visited the Yosemite. They were mostly men, and were admiring the beauties of nature, while forming court to a bright and fair New Englander. The latter's curiosity brought the whole posse upon us; the men loftily stared at the work; a couple of Englishmen condescendingly asked statistical questions. The girl pitied the sheep for a while, saying, "Poor things," and "What a shame!" But after watching a few come up feet first, sneezing and choking in the yellow soup, she could not prevent herself from a genuine pleasant laugh. It was good to see and hear; it was the last echo of softening influence which remained in our memory for six long months. The great want in the prairies is the right kind of girls.

Reverting to our uninteresting selves, we are all red-hot and weary, perspiring, dusty, and splashed with slops of the dip. All we can find time for is to wash our hands and faces, and the dinner is served: mutton-chops, bread, a can of tomatoes, and tea. It is not luxurious. There is hardly time for a quiet smoke when we see the dust of the afternoon band approaching; we must be off to work again; it must be done out of hand, for, as a rule, while the sheep are lingering round the dipping-station they

A SAGE HEN.

are poorly fed. The new arrivals are corralled, while those operated on in the morning are driven out to the feeding-ground. It is a hard lot, that of the sheep during this time of year; they are caught, sheared, dipped, and branded, necessitating a lot of driving, hustling, starving, and shutting in pens; they suffer under the process, and lose materially in condition. If the owner has a limited range, the sheep must be taken up into the hills for the summer; here is more tramping and travelling along dusty roads, with a burning sun and poor feed.

The afternoon's work is the same as in the morning, but the sheep put through are left in the enclosure all night, as by the time we have done work it is too late to take them out. They make a bad use of their opportunity by pushing the cover off one of the boxes, and tearing open some of the packages which contain the powder used in making the dip. Sheep will eat everything. Unfortunately, some fifty died of their experiment on this occasion. After having dipped the band they were all marked with a brand, and next day we started off along a by-road which skirted the foot of the hills, and finally rejoined the main road near the Merced bridge.

Driving sheep is simple enough in theory. The herd is marched from day to day a distance of eight or ten miles, feeding as they go, starting very early so as to travel in the cool, and, if possible, reaching the banks of a stream before the sun grows hot. Through the heat of the day the sheep do not care to feed or to travel; if full they will lie down, seeking some shade, or drooping their heads under the shadow of each other's bodies. This is called nooning; it may begin as early as eight o'clock in the height of summer, and last till four or five in the evening. It is a regular part of the day's business, and is often very troublesome, when you have a little distance yet to go, to find the sheep stopping in bunches, some lying down, and the whole baaing their protest against further exertion. If you want to reach your point now is the critical time; when the sheep baa shout at them and hustle them a bit with the dogs. Beware of a check, as should the flock once get bunched up your chances are over; you may then let the sheep lie, they will not travel again till evening. There is a disagreeable feeling of helplessness in handling sheep; they are the boss, and in your own interests you must study their whims.

Suppose, however, your arrangements have been good; you have brought the sheep to water; they have been pleased to approve of the quality and drink at once, without wandering off in search of something clearer, fresher, warmer, or different; it is not always we can understand their fancies; they will feed again for a little while, after which you may bunch them up where you can conveniently watch them. You will see some standing in a line, each head under the belly in front; others gather round a bush, with their heads together in the shade and tails out; some lie down to sleep; many stand with vacant eyes and noses stretched to the ground, and ease their feelings by heavy panting. In the afternoon, so soon as the sheep show a tendency to scatter out and feed, they are headed in the right direction, and travel slowly till nightfall, when they are rounded up in a bunch, and expected to sleep. A good driver will as much as possible fall in with the inclinations of the herd, and let them start, travel, and feed much as they are disposed, always, of course, with due regard to the prime necessity of getting over the ground. There are besides certain factors of which the sheep can scarcely be expected to be aware with regard to the situation of water and feed, and it will often be desirable to drive them even a couple of miles after they show a desire to noon, so as to reach water. Sometimes, to get across a desert, you may drive the sheep so much as twenty miles a day; but this has to be done at night if the weather is warm, and can seldom be ventured for more than two or three days. Crossing the mountains the sheep are often so much as four or five days on the snow without losing any large number of the band. After reaching good grass on the further side they soon recover themselves. At night the sheep, if well fed, will lie still, but as a rule when travelling they had better be watched.*

Leaving the main road was not on the whole a success. The feed was better, but on the county road we were more on our rights, and would have met with fewer annoyances than we had to encounter from small farmers, whose object often seemed merely to exhibit "cussedness," though, to the credit of the few, it must be said that their intention was elevated into the regions of common sense by the motive of extracting a few dollars.

---

* A cruel necessity was disposing of the newly born lambs. The mass of the sheep were yearlings; all were supposed to be dry sheep. New arrivals were *de trop*, and more likely to injure the ewes than be of any benefit themselves. There was nothing to be done but to knock the flicker of life out of the little things, and drive the mother on. The latter make no difficulty; at all times these merinos are careless parents during the first few days after the birth of their young ones.

At one place we were turned by a malicious falsehood along a road which was the subject of a local lawsuit. The proprietor was holding the way against all comers, particularly against sheep. We had known of this obstruction, and had arranged to make our way round; but the American sense of a joke was not to be balked. Fortunately the proprietor in question was a much better fellow than his opponents at law, and let my band through the disputed right of way. At another place we were amused by a woman running out of a farm and calling on her husband to "give them hell." These are little incidents, but will serve to illustrate the dislike American farmers have against sheep, and the petty annoyances they are not above putting in practice on the drovers. To see the worst side of the character of settlers in California, I could not suggest a better plan than moving a band of sheep through one or two counties; after that you may try anything else and enjoy the change.

The objective point of the drive was Sonora, which stands at the west end of the only road over the Sierra Nevada mountains which is possible by wagons in this part of California. There were several rivers to cross, where the only convenient points of crossing were farmed to some man who worked a ferry, and taxed sheep exorbitantly. The rates permitted by the charter often allow as much as fivepence a head for sheep and pigs; this the collectors themselves reduce to about a penny-halfpenny in their printed rates, but generally are satisfied with about half. Even these amounts, when they recur three or four times, together with the road-tolls, add a heavy percentage to the original cost of about eight shillings a head. It can only be avoided by crossing the mountains over out-of-the-way and difficult passes which are known to few people. The farmers who have lived many years near the hills, and have sent their flocks up regularly, hazard these passes, notwithstanding the risk of spending several days in the snow, rather than pay the heavy tolls. In this matter of tolls the Americans will not learn that three sixpences are better than a shilling.

The season was fairly early for travelling; the road had not been cut up by much stock going along it. The sheep did very well, getting enough to eat while advancing about eight miles a day. 'The last day's march up to Sonora was much longer than it should have been. The boys pushed on so as to reach the town, and revel for a night in the charms of a saloon.

If the love of drink is to be regretted, which enslaves the lowest classes of towns where these are brutalized by crowding, poverty, and misery, it is, I think, sadder to note the hold whiskey has on the Western men who, from education and surroundings, rank in intelligence far above the working classes of the cities, and to whom prosperity and comfort are within arm's-reach if they would only let go the glass to seize them. The weakness for whiskey is universal. On ranches or in camp it is never kept, the result of an opposite course being certain; consequently so long as a man is at work he perforce abstains from liquor; but the moment he is free, or should he pass within a certain distance of a saloon, the temptation is too strong; his craving impels him to neglect work, or to undertake a toilsome journey to obtain the exciting cordial—but poison one would rather say, viewing both the quality of the mixture sold generally in small saloons, and the constant excess to which it is indulged. There is not a man but thinks of it, and pines for it, and so soon as he has an opportunity will indulge; nearly always he will take too much if he is employed where it is not within daily reach.

Of the men I engaged some were very fairly educated, and in the ups and downs of colonial life had seen varied fortunes and tried many trades; but with the exception of two men, who had been raised in severely temperance States, and retained partially the habits of their education, all the rest were devotees of the bottle, and most of them got drunk. Cooking is not a special trade on the prairies, where all men learn to cook by having to depend on themselves for a meal; not everybody cares to undertake the job, but the man who does is just the same as his fellows with whom before or hereafter he may be working or riding; for the time, however, he has a specialty, and is somewhat of a despot. What is the connection between cooking and drinking? No one knows. I employed four cooks; they all got drunk when fortune favored them. Two of them were by far the best men in the outfit—quick, clean, industrious, early risers, and fairly successful in their art—they were clever drivers and looked after their horses; but given the time and the money, the result was an easy prophecy. Whenever I heard of a saloon on the road I tried to arrange the drive so that we should halt as far from it as possible; the boys would grumble a bit, and the more thirsty would start at eight or nine o'clock at night to walk back a couple of miles; their return was invariably noisy.

Light beer, brewed by the Germans, who have flocked

by so many thousands into America, has diverted in the Western States the taste from whiskey. It is, unfortunately, too bulky to be carried far off from the railway, and the breweries are few out West, so that the rail-carriage would be expensive. Californian wine may hereafter compete with whiskey. At present hotel-keepers prefer to keep native wine on their list at extravagant prices to selling much; good wine can be bought for from three shillings a gallon; it is generally charged at from six to twelve a bottle in the small towns. In Southern California, where wine is cheap, a pleasant light wine is supplied sometimes gratis with dinner and supper. I did not notice any Americans drink it. Mexicans and Southern Europeans were the only ones who seemed to care for it; yet from an old-world prejudiced point of view it must be far more wholesome than a hot cup of strong tea or coffee, taken with meat. As a matter of taste one must say nothing.

From Sonora onwards, except for a few miles at the beginning, the road runs through the forest, and is quite unfenced. This is about the most difficult part of the drive on account of the loss from sheep straying into the bush; generally a few extra hands are hired—often as not Indians. The latter belong to the Digger tribe, and some of them are not averse to work either on farms or in the town. They are not all equally civilized, and one of their little settlements of a few miserable hovels, with granaries of pine-nuts in the shape of beehives four feet high, enclosed by a poor fence made of brambles, cut down and thrown into a line, gives a notion of their aboriginal and miserable style of living. The picture will be completed by supposing an ancient and wrinkled hag sitting on a flat rock in the ground pounding the pine-nuts into flour, the mortar being a hole in the rock itself. For a few marches out there are corrals, in which the sheep can be placed at night, and out of which they can be counted in the morning. This, however, takes so long a time that, as a rule, it is done only every second or third day; counting the black sheep, and those with bells, being thought a sufficient check for intermediate occasions. It is quite possible for a bunch of four or five hundred to disappear out of a band of as many thousand, and the ordinary herder will not notice their absence, even in an open country where he can see his flock together.

Crossing the Sierras, a very small portion of the band travel on the road. Most of the sheep are scrambling along the hillside in a parallel direction, browsing on the young shoots, or wildly climbing in search of young grass. With all this bush to contend with, it was hard work to keep the sheep together, and it is no unusual sight to see a band, as if gone mad, mounting and mounting towards the hill-top, scattered everywhere in groups of ten to twenty, striving to out-run or out-climb some bunch with a slight advance, baaing and rushing as if distraught; and all because they have come on a patch of wild leek or green snowbush, butter-weed, or brier. Now is the occasion for the shepherds to show their activity; they must outpace the sheep in climbing the hill, and strive to turn them in fifty places, or they will have a small chance of collecting the rabble without great loss. A dog in such moments is of more use than three men; not only that he gets more quickly over the ground, but that the sheep mind a dog, whereas they have no fear of the men. When started on one of these escapades, they will stand and dodge a herder, or turn only so long as he is driving them. Others would sneak into the bushes, or hide in some little ravine, while nature aids the troublesome brutes in exhausting the men, who are often taken in by the appearance of rocks far above them, and thinking to catch a band of strays don't find out their mistake until they have had a long climb.

We were always glad enough when night came, and could see the camp-fire alight ahead. Towards evening the sheep follow well; it would be as difficult to separate them now as in the day-time it was hard to bring them together. No longer in search of food, they come down to the path, succeeding each other in endless line. For a quarter of a mile the road is a solid mass of woolly heads and backs, with wisps joining in at intervals from out of the dusk through some gap in the bushes, or down a broken ramp in the bank.

A bedding-ground has been chosen already, and so soon as the leaders reach the further limit they are stopped; the rest crowd in, and are made to close up their ranks; the men and dogs walk round and check the usual discontented ones who now want to go foraging. There is plenty of dead wood, and soon half-a-dozen fires blaze at various points, lighting a small portion of the forest, and picking out, with a ruddy glare, the outlines of the men and pine-trees. By and by cook shouts "Supper!" One man is left on guard, and we gather round the piece of oil-cloth spread on the ground, on which are laid the exact number of tin-plates, etc. After supper the watch is settled for the night; we all turn in except cook, who is left washing up and getting everything ready for the speediest preparation of breakfast next morning.

There is little game near the road. It was not till close to the snow that we saw a few deer; there were some grouse, and on the tops of the hills we occasionally came on the traces of bear; this does not imply that there is no sport, but to find game it must be sought. The traveller must not expect to pick up a few head while sticking to the road. Just in this portion of the Sierras there have been no groves found of the gigantic Sequoia, although two of the best-known clumps lie at no great distance north and south of Sonora. There are, however, magnificent sugar-pines but little inferior to the Sequoia; in reality, a handsomer tree, and one which stands grandly on some of the boldest mountain sites. The streams at this time of year are full, and refresh us with a sparkling draught or a bracing bath under the noonday sun. The road follows the valley of one branch of the Stanislaus River, which, as we see it, is a dull green ribbon far down below us. The bigger streams crossing the road have been bridged; some yield trout; but for fishing you must wait till they are less turbid.

After ten days' travel through the mountains the herders were pretty well tired out by the unwonted exercise of chasing vagrant and skittish yearlings along the steep and rocky slopes, or in slowly pushing their way in rear of a straggling bunch through a labyrinth of tangled manzanita or bull-brush. Here you have to contend each step with the tough branches, forcing the upper ones apart with your arms, while you feel with your feet for some firm footing in a mixture of low ground-stems, roots, and loosely holding stones. It is bad enough to work your way down-hill; but if you have to mount upwards with a band of a hundred sheep to watch, and bring them back to the road; to head off those which foolishly fancy an outlet by some small clearance to one side; to keep the leaders in view and in the right direction; to persuade those lagging behind to follow at all—you will enjoy no small trial of your calf muscles, and a moral victory if you repress the bitter anathemas on the whole race of sheep. It is no dashing occupation that of sheep-driving; it requires endless patience.

At last the band was gathered near the foot of the pass. One bunch had strayed, where and when was not quite certain; the boys think they must have got away up the mountain; there will be time to go back and hunt for them before the wagon can cross. Feed is scarce on this side; the best thing is to get the band over as quickly as possible. There remains a steep road over barren hills, which mounts continually upwards for six or seven miles before we reach the snow. The young wild-leek grass which had made the sheep so rebellious had not sprouted at this higher elevation; the snow-brush had put forth young shoots in warm corners facing the south; the long hours of sunshine and heat-conserving rocks have forced on the vegetation of such bushes as the wild rose and flowering currant. There was about enough to feed the sheep, just for the day, up to the verge of snow, but we could not afford to halt, the principal aim being to take the sheep over at once while their strength is good.

One band, belonging to a well-experienced Frenchman, has crossed the mountain already just ahead of us. The difficulty of the passage lies in the danger of falling between two seasons. Up to a certain time in the early summer the nights are sufficiently frosty to harden the surface of the snow, upon which the sheep can then travel without floundering, to about ten o'clock in the day. A month later the snow will have disappeared enough, not only for stock, but for all ordinary wheel-traffic to follow the road. At present, the nights are too mild to please us. The hot sun during the long days sends the smaller torrents roaring and splashing across the road; the sheep hesitate to wade, and hunt for places where they can cross by leaping from rocks to shore, until a certain number are over, when the rest disregard a wetting, and rush through quite boldly.

Pushing on ahead, I find the snow a little soft, but still passable; it promises, however, to be hard work for the pack-horses. We must get on, there are many long stages ahead, and every day's delay makes the further journey more difficult. In the mean time, report tells us of several other herds echeloned along the road, who also intend to cross. We must keep our bands clear. Should they get mixed, it will take days, even after reaching a corral, to separate them, and in the process do the sheep a great deal of harm by the handling. At the end of the seven miles the road crosses a creek, and on the opposite shore the snow begins; it covers the road for five miles to the summit, and for about three miles on the further slope. The sheep are particularly stubborn at this trying moment, and cannot be made to ford the torrent. With each hour the water rises somewhat, but it never becomes a serious obstacle; it is a case of will not, which occurs among sheep, and is most annoying, as you feel powerless to control the inert resistance of the crowd of stupid animals; to get a right movement out of them seems a hopeless task. Valuable time is wasted; not only the sheep have not crossed, but standing bunched up they necessarily have lost the chance of feeding for that day.

We survey the creek for some distance up to find some better ford, but without success. There is no chance of improvement, except at one place, where in former years a bridge had spanned a narrow part of the stream. The poles had fallen off the natural rock on which they had abutted on one side; but the snow partially overhung the chasm, and if this held good we might repair damages sufficiently to make a crossing. The men could cross easily enough by fording, though the water running freshly from the snow-drifts above was icy cold and chilled our legs, which was not an attractive introduction to standing in the sloppy snow on the bank. The work was urgent; a pine-tree is cut down, but before we can get it into position the torrent catches the spread of top branches, and wrenches the pole out of our hands; we see it merrily plunging through a series of small cataracts. A couple more trees are felled, and, being more carefully handled, they are laid successfully across between the fragment of the old bridge and the snow on the opposite side; branches are thrown on to stop holes and make a platform; by night we have secured a passage which should be good enough.

The sheep were driven next morning to this bridge, but were listless, and disinclined to face what was scarcely a difficulty. In addition, the boys were tired and apathetic; matters had not gone quite smoothly; therefore they were disheartened and grumbling, and failed altogether in the average pluck of white men. Fortunately, the owner of the herd immediately behind had come on ahead to see for himself how far the coast was clear; he was equally interested with myself in getting the band over, and lent a willing hand. The sheep were driven up so closely to the bridge as they could stand, but of course met our efforts by continually breaking away, clambering the rocks, and dodging the men who stood round. A certain number had to be caught and dragged by a leg along the snow and over the bridge, in this way making a trail; and by standing on the further shore suggesting to the band that remained behind the object they should attain. At last one sheep more bold leads the way, walking with hesitating steps, and sniffing the air with head erect; he seems to question the possibility of the small bunch standing out there in the snow being really companions; he crosses without ever looking at the bridge. This is all that is required to give a start. We need only be still, and not risk frightening those which have passed over. By degrees others follow, at first slowly, but soon as thickly as the width of the bridge will allow. We must keep the whole band well pressed up against the stream, or the line may be broken, and the passage would be interrupted, and maybe have to be reopened by further catching and dragging.

We happily avoid any serious difficulty. The line is not badly broken; half the sheep have still to cross; the other half, which at first had stood huddled in the snow, have suddenly taken it into their heads to climb, and in a few minutes they are off. Selecting the faces of the slopes where the snow lies thin, or the soil is exposed in patches, without any prospect of feed, they are scampering like chamois. These must be stopped; it takes half the men to do it, and that not without hard toil.

I must go back to camp for a time, and arrange what articles are indispensable for the next fortnight. These will have to be packed on ponies, while the balance can remain in the wagon, which, with the horses, will be left at the foot of the pass until the road is open. It is a hard day's work; the weather has changed, and a slight drizzle has succeeded to the days of sunshine. Late in the evening the ponies are packed, and taken up the river to camp. We must ford the stream, and, after unloading, the ponies have to be brought back again, as the only chance of their getting anything to eat is to tether them on the further bank, where, if hungry enough, they will browse on the bark of the willows. They get a small feed of barley; we cannot afford much, as the steepness of the road had limited our stock. Having picketed the ponies we return to supper. It is very late; wet to the fork in crossing the icy stream, standing in slushy snow, we eat our meal round the best blaze we can make in a clump of pine trees. The sheep are scattered irregularly along such portions of the road and bank which are clear of snow.

Next morning the pass must be crossed. It takes some time before the sheep can be induced to travel through the snow. At length they are off. The four ponies are packed; there is nothing to do but to follow the trail of the former band, and reach the other side, which is not more than a day's journey. Unluckily at this point I was obliged to go back, partly on account of business which had been carefully muddled in a combination of bankers with the express (parcel) service, and partly to hunt up any strays which might be found. The band was put in charge of the fittest man on the spot. There was nothing to do but to cross over and keep the sheep in the best condition until I could rejoin.

The trip back to Sonora did not take long. I arrange my affairs, and, having got the wagon with one of the men, I load it with a good supply of provisions, knowing that the commonest articles are of extravagant price when one you reach Nevada, and start again over the same road. As I have to hunt up the stray sheep we add to our number a couple of Indians, who are thought good trackers. Although we have to walk a great deal, high upon the hills, hunting for signs, it is on the whole pleasanter than being at the tail of the herd. The country is quite wild. Along the first thirty miles a few shanties are inhabited, but ten yards on either side of the road the forest begins, and the further you leave the road the wilder and more solitary becomes the scene. There are no animals, and few birds; except the one road there are no paths. The higher parts of the hills are fairly clear, but now and again the vegetation along a stream is so tangled that to get ahead you must go a long way round. Signs of our sheep we cannot find. I hear of two bands which have come in since I passed through, and as small bunches of sheep will always run into a herd, perhaps I may yet discover some of the vagrants.

On one occasion I took one of the Indians along as more likely than myself to hit on the trail, and follow it up to the band for which I was searching. We did not have much trouble; while making our way into a valley we heard bells, and not long after the flock appeared, scattered, and running like deer to feed on the young grass. As the herders must be following we sit down on a prominent point, and have not long to wait. Two men are coming up the valley. In the hand of one of them something flashes at intervals as he swings his arms. The Indian nudges me and says, "Man got a gun." True enough; the herders were out watching their flock, each armed with a revolver. They were not, however, wild spirits, but steady men. There is sometimes a misunderstanding about ranges; two different men will claim the same one. The story went that some Frenchmen had murdered a man just hereabouts, and driven his band of sheep over a precipice. It was well to be ready to protect yourself. In the West there is little law to safeguard property; none for life. It is, after all, but rare that business takes the serious turn of shooting. Most of the frays rise out of gambling and drunken quarrels, and shooting is relegated to the saloons and haunts of the most depraved. My Indian was a cautious man, and kept out of sight until he saw that it was a friendly pow-wow. Not that his particular people are much ill-used; at present they are few in number, peaceable, and have no large reservation to stir the envy and covetousness of the white man.

The other herd had gone off in quite a contrary direction with a view of making its way over the mountains by a less-used pass; where there was no road, consequently no toll. It took a long morning to follow the trail and reach the herd, which had followed up the brow of a salient spur; in neither case did I find or hear of any of my strays. The idea is that sheep left to themselves will mount to the highest points and remain there unless disturbed by bears, wolves, and cayotes, which will split up the bunch and scatter them. On this view of what the sheep ought to have done, I hunted the hill-sides above the road, and sent the Indians in other directions. After six days' toil we reached again the foot of the pass, and the Indians were sent back home. The sheep were subsequently found far down the valley.

There is a heavy expense in taking horses through the mountains, for not only is barley expensive, but, as there is little grazing, hay has to be freighted out to the different points, and varies from one-and-a-half to three-and-a-half cents a pound. When it comes to feeding big horse, thirty or forty pounds do not go far. I receive a letter from my temporary foreman, sent by the hand of a traveller who had just crossed, saying that he has hired a range, beside entering on other transactions, and asks for a big sum of money. This is a serious business; if I will only give him time he will, I feel sure, like an electioneering agent, study my interests by getting rid of any amount of money with the greatest industry. I determine, therefore, to leave the wagon at the foot of the pass, and to ride over the team-horses.

There is a great change since the sheep started a week before; another mile of the road is clear, but beyond snow still covers the country, though large patches of ground are bare on the south-facing flank of the pass. The tracks of the sheep which have crossed show but faintly, as the surface of the snow melts off each day; it it is impossible to keep the road, of which the direction is at times difficult to trace. It probably would follow the bottom of the valley, in which the snow still lies three or four feet deep, and is so soft and rotten that it will not carry a man, much less support a horse. I try to get round by the bare spots, lead the horse, and carefully feel in front of him when about to cross deep snow. At last I come to grief—the horse breaks through the crust, and is buried all but his head in a snowdrift. He has tumbled over the concealed trunk of a fallen tree, under which his legs have slid, while his body is on a slope which tends to keep him fast. One hind-leg is hooked up in an awkward position. I take off the saddle and luggage, and encourage him to extricate himself. It is of no use; and fearing lest he should cut his legs with his shoes (for much snow-work the shoes are always taken off), I leave him, and run back half-a-mile to meet some Portuguese, who are bringing over a herd of sheep. From them I borrow an axe and an iron basin, and, with the help of the man who is coming on the other horse, we dig and scrape a bit, and the horse releases himself, plunging down-hill through the snow. It is evidently not much use to try and take the horses over, so I send them both back with the man, and set off on foot.

Of the Portuguese, not one speaks English, and only one knows the road. He is occupied in forcing a small bunch ahead to serve as leaders for the rest of the band; he understands my questions, but has his hands full, and can give no answer but by waving his arms upward. It's small fun starting late to cross a summit without a road, not knowing which valley is the correct one, and having to wade through soft snow half the time. There are two valleys in front, either sufficiently large to be followed by a road. Of course with ample time there is little real difficulty. You hunt the tracks of travellers who have preceded you; but while going on roundabout ways you lose the trail, and then grow impatient; you begin to fancy that you ought to have taken another valley; you think you noticed a blaze on a tree in that direction. You do not want to go a mile out of your way with the double dose of returning; and then the chance of stopping out all night without food or blankets is not inviting.

This is mere nervousness, but still uncomfortable. It takes time for one whose senses have degenerated through civilization to be quite at his ease in the midst of a boundless solitude; and however ready the newcomer may be to do as those around him, the more acute sensibility and vivid imagination bred of a complex civilization is a burden which he will find difficult to throw off. One effort will not do it; it must go in pieces, as experience shows that man wants but little, or rather that when he has but little he can get on very well without the rest. I must push on, as even after reaching the road on the further side there are some fourteen miles or so before reaching shelter.

Looking back on one's little expeditions and trials, they diminish so dreadfully. It is most discouraging to travel nowadays, after reading the journals of a century back, when you need not leave Europe to find excitement and hair-breadth escapes from robbers, sinking beds, haunted castles, and barbers' shops fitted with underground cellars and sausage-machines. Now you lose your way, you find it. You have missed your lunch, and therefore eat a heartier dinner; nothing is simpler, even in the by-ways. I take the right-hand valley, and occasionally satisfied by the hoof-marks of the last pack-ponies which crossed, I arrive at the crest of the pass. After this it is plain sailing. The snow on the eastern side of the Sierras reaches only about a mile; the road is plainly visible some distance ahead; the hills below look green; and far off a small lake is twinkling like a burnished silver plate. It is all downhill, but the sun blazes hot; there are hardly any trees. I find a cabin which is occupied, but no one is at home. I have a crust of bread, there is no want of water; with these I make a lunch, and enjoy a short siesta.

About three o'clock I start again, and after nine miles more walking, including a good deal of wading, as most of the streams follow their own route, either across or along the road, I rejoin the herd, which is camped on both banks of the Little Walker River about half a mile from the main road. A sheep-bridge connects the two banks, and corrals have been roughly made by chopping down young cottonwoods. There are many sheep-bridges all over these ranges, made by the people who have herded here, which are well known only among sheep-men; if you are aware of such a construction, the converging sheep-paths will readily point it out. It often consists of a single large pine-tree, which stood conveniently on the bank, which has been felled, and directed in its fall across the stream; sometimes a little labor is bestowed in flattening or cutting notches on the top surface. If often to be used, a rough balustrade is added, and a few stones are piled to make a ramp by which the sheep can mount readily on to the log. The banks of the river on either side, above and below the bridge, may be fenced, to prevent the sheep from pushing each other into the water when crowding to cross. Over such bridges sheep make no difficulty in passing; the smell left by former bands give them confidence. Owners whose bands are constantly travelling train one or two stout wethers or goats to lead the herd; the latter, I understand, have to taste the stick before they thoroughly imbibe their education; they are great tyrants in the herd, and mar their usefulness at times by leading the sheep into difficult places where they are liable to fall and injure themselves.

I find the sheep in good enough condition. They have nothing to do but to feed on the fresh grass, with plenty to drink, and no distance to travel. The younger ones play and butt each other, dancing and bounding in the air, alighting sideways, kicking up their heels, in the most graceful and lively manner. I hear that they had a hard time in crossing the mountain. The passage was most clumsily mismanaged. The whole outfit had spent a night on the snow. The two men in charge of the pack-ponies had found some difficulty in getting along; their hearts failed them; so, pulling the packs off, they left the ponies, and, blindly deserting food and blankets, they went off to rejoin the men with the herd. The whole lot spent an uncomfortable night, and next morning, demoralized by the idea of reaching a saloon, they abandoned horses and packs, and drove on the sheep. At the first shed they were disappointed as to whiskey, but bought some food. Hope told a tale of another hotel beyond, so they rush on again. The foreman, delighted with his rise, goes ahead, hires a horse, and indulges a spirit of benevolence to all humanity at my expense in buying food, lodging the men, hiring Indians to recover the horses and goods left on the snow. Three of the horses are brought back in miserable condition; the fourth died. Some of the men have fallen out with the foreman, and want to quit.

In another way one or two of us have been entrapped, when paying the toll on the Sonora side. We understood that no further toll need be paid on this side of the mountains; but as the last ten or eleven miles of the road are in another county, this county has a right to grant a separate charter, which it has accordingly done within the last ten days. A friend gets the monopoly; he employs half-a-dozen men for a fortnight after the melting of the snow to clear off big stones, from which time he has a snug little income. The collector misses the first band over, but I am not so lucky. He and the foreman, whom he afterwards told me was a perfect gentleman, agree in an off-hand way that the toll shall be nothing more than just enough to throw the stones off the road which the herd will roll down; they trade a horse for some sheep, arrange for some grub, and behave with the handsome disregard of expense which animates us all when handling other person's money. I need hardly say that the bargain on my side was a bad one.

Sheep are looked upon as natural prey by all outsiders, and it is a race who can take most out of the owners by hook or by crook. There were many stories current on the California side, of ranchers setting wire nooses and other traps in the bushes to catch passing sheep; these are mere stories possibly, but I was told by the boys that in crowding through a village they stopped a man from pulling one of the sheep into his doorway. A single sheep is of no great value, and such a proceeding would be held to partake of the nature of a farce.

Next day, after arrival, attention must be paid to the quarrel among the men; they cannot get on together; some of the younger fellows wish to quit if I cannot re-model the service to suit. They had left the half-band of which they had charge for a whole night scattered on the hillside, and in the morning had gone leisurely to work at nine o'clock to drive them together. They objected to bringing them to a corral where they can be counted, and had behaved in very poor form; the only thing left was to part company.

The team-horses followed me across the pass two or three days later, and using these I rode all over the hills for a week to assure myself that no more strays were lying out. Notwithstanding very circumstantial reports of sheep having been seen in certain places, I could find no traces, and subsequently came to the conclusion that the number said to have been counted in the band was deliberately mis-stated. Why one of the boys, who followed our fortunes for some two months, persisted in supporting the story when it was of no disadvantage to anybody to tell the truth, was difficult to understand. It was a small amusement to me, when I had fathomed the scheme, to hear his views as to where the mythical sheep had gone to. He might as well at the time have saved me and the horses many days' hard work, much annoyance, besides some professional disgrace which attends a herder who loses stock and cannot recover them. The ranchers around, getting hold of the story, congratulated themselves on the whole band having been scattered to hell by the cowboys. Such is the affection that men who own cattle bear to men who own sheep! The traditional belief is that, once sheep have crossed a range, cattle and horses leave it. There is a foundation of truth: the larger stock will not remain on the same range where sheep are regularly herded. The smell left by sheep is very strong, and, I suppose offensive. But, strangely enough, pigs—which to my nose smell worse—get on very well with horses. With such objections, the further sheep keep away the better pleased farmers are. In some counties the latter have passed laws prohibiting sheep being herded within ten miles, or some such limit, of a dwelling.

The discord in camp having been settled left me short-handed. The young fellows went off to try for the more congenial occupation of cow-punching. They had one pistol, and a growing ambition to distinguish themselves by breaking the law somehow; their lot fell on hard places, as soon after I received letters from three of them offering to rejoin. But I wished to hire older men, who had experience in sheep-driving, and, if possible, knowledge of the road we had to travel. The only chance of hiring help was at a mining-town some thirty miles off, and for this place I started with the foreman, reaching town on a Saturday evening.

Next day we found two men, one of whom had had practice with sheep; the other was one of the usual all-round men, who considered that he could do most things, and vaunted his precise knowledge of the trails throughout the territories. "No one could fool him on the road!" This was so far promising; we had only to rejoin the herd, get the wagon over the mountain, and make a start. Previous to my leaving the town, the foreman, who had discovered an old friend, asked leave to stop behind for a day, and refresh memories of old times. I was not at all averse to his having a good time before grappling with the drudgery that was before us, and sealed my consent by a necessary advance of coin. One of the new-found men accompanied me in the buggy down the hill; the other with the foreman, would follow on horseback the next day. The new man accordingly turns up, but the foreman has stayed behind, and rumor says that my friend was last seen in high feather, and unwilling to leave his paradise. After a few hours a telegram was put into my hand "to come at once or send money." As neither alternative was to my mind, and either would have been an absolute waste, I declined.

The details of my friend's adventures were the subject of a humorous article in the local paper. The story commenced with the arrival of a gentleman with his pet herder, who is described of a festive and light-hearted disposition, foreign to the conventional type of man who follows the melancholy calling of tending sheep. Some business having been transacted, the former left, probably with the intention of being present at divine service elsewhere, while the festive herder remained to delight and astonish the citizens. Having invested in new clothes and some whiskey, this philosopher, who gathered roses within his grasp, exalted himself in his own estimation, first to the position of partner in 5000 sheep, and finally to the complete ownership of 10,000. Notwithstanding that on first arrival at the hotel, in reply to some question, I had mentioned that this man was my foreman, he was clever enough, under the inspiration of good spirits, to induce several people to advance him money. On this fund he entertained a choice circle of friends, old and new, and caroused for a couple of days. Not content with this game, he next invited two ladies to join his fortunes, and accompany him and his herd to Montana, where they would all live happy, healthy, and wealthy. He explained to his fair friends that it was somehow necessary to assume the disguise of herders, to which they agreed; one wearing her own short hair, the other consenting to sacrifice her locks to the situation. The following morning this Don Juan mounted his two candidates for employment on the horse which had been left to bring him down, and well primed they started down-hill, leaving his debts behind them. Unfortunately for the termination of this comedy, doubts had fermented in the minds of those who had parted with their money; the law was put in action; a charge of obtaining money on false pretences was sworn against him, and the sheriff started in pursuit. About five miles out he found the three pilgrims in high good humor, and regretfully interrupted the hopeful cavalcade by presenting his warrant. Our hero was not daunted; finding a soft stone, as the local edition asserted, he left his two companions, recommending a little patience and trust; he promised to settle matters shortly, and to relieve them speedily. In town he was I believe, unable to find bail, and was locked up. The disguised ladies must have looked in vain for the dust of his coming, and finally reckoned they would return to town. The story of their appareling and adventures had got abroad, so that when the grip of the law on their protector and patron was added to the programme, the *denouement* was anticipated; times being bad, and the town full of idlers, the citizens were able to spare leisure to give the distressed damsels an ovation, and present their cordial sympathies in the shape of a comic welcome. Threats of dire revenge were breathed against the deluder, who, however, was a man to compel fortune; he satisfied his creditors; the case was not pressed, and with an undiminished fund of assurance he returned to camp a much injured man, and indignantly taxed me with so readily withdrawing my confidence. His anger was, however, moderate; he really was not a bad fellow in the main; a chance of a good salary, and promotion over other men, turned his head, and his vanity was accentuated by whiskey. His powers of persuasion must have been extraordinary to take in a lot of sensible tradesmen who knew nothing of him, to confirm his representations, and to induce two of the unromantic sex to lay aside rivalry, and cast in their lot with his. He succeeded in generally pacifying his creditors, and in camp borrowed a month's wages from one of the boys, who not only knew a little of him, but was also up to the incidents of the last jink. With me he squared up all right. He had been a spendthrift with my money, but had not misappropriated any; as he put it, he was a white man, and so he was, and is so still, I hope. The end of the story was another article in the newspaper, which I was afterwards told rehabilitated me, and defended my reputation from unfriendly critics as having any share in the farce. What might have been the last scenes without the interruption which prevented the joyous company reaching camp, was amusing to contemplate.

At length the sun and the weather did by themselves what the collectors of the tolls would do nothing to expedite—that is, clear the road over the pass for wheel-traffic. With no very great expenditure a way might have been opened ten days earlier, but with a monopoly the main business is to receive; to give an equivalent is not so important. We rode over the two team-horses; the road appeared passable. We met a wagon going to try it the afternoon we arrived, and next morning hitched up our team. With a little skirting round snow-drifts, and floundering through sodden ground, we reached the top of the pass. The road down was all right except in one place, where deep snow lying in an elbow of the road necessitated a diversion. The cut-off was a sharp descent; the hind-wheels were rough-locked, —that is, a large-linked chain was tied round the rim of the wheel in such a way that the wheel rides upon the chain, which drags along and cuts into the ground; the other wheel is jammed tight with the brake, the horses are turned off the road, and the wagon, though nearly empty, slides down the hill in great style. From this point there was no further trouble all the way to camp.

Everything is now together. On the 23d June we make ready to start. The stock of food required very little to be added to it: the wagon carries a ten-gallon barrel for water on either side; we have flour, bacon, syrup, beans for over a month. The price of commodities in Nevada will not affect us for some time; for meat there is the flock. It is no small job to catch, kill, and skin a sheep every third day. Luckily one of the men is an expert, for the boys make a study of putting work on other shoulders. They think it is the cook's job; the cook on the other hand says, "You give me the meat, I cook it."

In this corner of California there is a network of toll-roads which, running in different counties, are parcelled out in short lengths. Now, as roads are quite unnecessary for driving sheep, being in fact rather a disadvantage, and the tolls exceedingly heavy, it was a prime necessity to avoid them as much as possible. Further on there was a strip of waterless country, called the desert, which we had to cross, and at which point to do this with least harm was also a serious question. The roads of course take advantage of the best gaps in hill ranges, and of the easiest river crossings; and what with a rocky pass or a bridge they generally succeed in compelling all traffic to employ them at some point. The possibility of stock being able to travel along the hills is checked by the cross-fences which farmers run over the Government land near their ranches. Fortunately one of the men has lived for some time in the locality, and he manages to steer us clear of all toll-gates. The way is crooked, but so much the better, that in avoiding towns the sheep find good grazing.

It is no sinecure to manage a band of sheep; so many things are required, while the distance within which they must be found is limited by the short daily journeys which the sheep can travel. So far it has been fairly easy to find grazing and water, but after entering

Nevada our troubles began. You must manage to drive the sheep where they can feed as they move along in the right direction. Before noon they must have water. At this camping-ground the horses will require grass, which they cannot pick up on the scant tufts which satisfy sheep. In the afternoon the same programme for the sheep; and for the bedding-ground a bare open spot, dry, free from feed, and away from damp; on the other hand, the horses want grass, and the camp wants both water and fuel.

At times the wagon cannot follow immediately on the tracks of the herd, but while these take a short cut the wagon has to go round. Here is an opening for some one to go wrong. The country now is new to all of us; the road is often merely a trail which splits and divides. There is no one to help you in a fix—you must judge for yourself. It occurs, therefore, that the wagon sometimes (not often) goes astray; and having lodged the herd in a place of safety, the boss must go off and hunt the wagon; or if piloting the wagon he must hunt the sheep. People who have not tried it cannot understand the difficulty of seeing at a distance even so many as five thousand sheep unless they are close together, and raising a volume of dust; but we are tolerably lucky, and generally manage to hit it off; and, though a few times late, we got our dinner and supper regularly, and did not have to sleep out of our beds more than twice on the whole journey.

I was surprised to find how little the people hereabouts seemed to know about the country roads and trails, even within twenty miles of their homes. I could not find any one who had crossed the desert, or who had any sound information on the subject; whereas in the more newly settled territories, as Wyoming and Montana, the boys seemed to be at home, so far as knowing the roads, over the whole place. Knowledge of the country roads has decreased considerably wherever the railway has supplanted freighting. The freighters are a special class, who have much to tell of the glories of the road in the days gone by, and of the money to be made in carrying goods from the terminus of the railway to the mining towns, which were in the full swing of prosperity. The freighting was done either with ox-teams, or horses and mules. The former moved in ponderous convoys in charge of a boss; the men, nicknamed "bull-whackers," travelled on foot. This profession never reached any degree of eminence; on the contrary, the man who had charge of twenty-four horses or mules, drawing three heavily laden wagons, one behind the other, was a responsible person, and knew his value. The property in his trust was considerable.

It was really a grand sight to see a string of well-fed mules with their massive harness coupled to the long-drag chain. The driver is seated on the near-wheeler, with leather leggings on, and a broad-brimmed hat; his whip hangs round his neck; he composedly rolls a cigarette while he watches that each mule does his share of work. The leaders are well trained, particularly the near one; to his bit a check-line is attached, which travels back over all the near mules to the driver. A steady pull turns the trained mule to the left, while two slight twitches turns him right. A light iron bar connects his collar to the bit of his companion, who is thus guided; the others all follow; the wheelers are saved to steer the wagon past ruts and hard places: the team with short steps develop great power, and work admirably together.

The front wagon is enormously heavy and strong, and carries a proportionate load. If not more than two others follow they are also of extra size; they are connected each to the hind-axle of the wagon; in front by a short pole. The brakes are very large, and can be worked by a rope at the end of a lever, which comes forward to within reach of the driver's hand. Should the road be difficult or hilly an extra man is carried, whose duty on the journey is to look after the brakes.

There is an enormous saving of human labor in this system of freighting. It requires a man of resolution, readiness, and self-reliance to tackle it. The roads at all times are bad; the more they are travelled the worse they become; the wheels cut deep ruts till the bed of the wagon drags on the natural surface of the ground, and a new trail must be picked out, which in the absence of any engineering is a hazardous experiment. In the winter, when the snow covers the bad places, the perils increase, and a sidling bit of ground, with a coating of ice, threatens the whole convoy with a swift descent to Avernus, which can sometimes only be avoided by a quick wrench up-hill of the wheelers in order to overturn the wagon, and so save the team and goods.

In the West there is no such thing as a made road; at best there are a few bridges near the towns. Some one who knows the country first ventures with a wagon, probably out hunting stock; the wheel-tracks direct others; by degrees a defined trail is formed. This trail is continually liable to interruption; either cut across by small ravines, or the melting of the snow in the spring send a stream of water along the deeply scored trail, thus entirely destroying it; the wagons have to hunt a road to one side. In driving along a strange road you must be continually on the alert to notice if fresh wheel-tracks break out to a side, and to guess the reason thereof; the general advice is to follow the last wheel-tracks, otherwise you may be following an old trail which ends in a tight place. Often you will find yourself on an apparently unnecessary loop; the track is right square ahead, but by-and-by you pass the skin and bones of an ox or horse which had died in its tracks; it never occurred to the owner to drag it off the path, and in consequence all subsequent travellers have had to leave the road and clear a new line through the bush. The most tiresome driving is through sage-bush; this is often strong, and holds solidly; it requires good steering, and a hard-working team.

After passing north of Aurora, for a short time a flourishing mining town, now the most desolate-looking collection of deserted houses and dejected inhabitants, we follow an old road to Belleville. The scenery is varied and interesting as we climb small ranges, but water is scarce, and we begin to experience the bother of dry camps and a thirsty herd; we are lucky in that we have enough water for ourselves; found sometimes in a hole or well in which it is green with alkali, or in an out-of-the-way spring, where a little clear water rises and runs for a few yards. It takes an hour to ladle up enough to fill our casks, and give the horses a drink. From midday on the 28th the sheep got no water till they strike a stream late in the evening of the 30th, when they rush in and convert the quiet willow-lined ravine into a very pandemonium, running up and down the valley, baaing, trampling the young grass which we are trying to save, at one end, for the horses, gorging themselves with willow leaves, and striving to break away in a dozen places,—manœuvres with which we have to contend in the dark; rolling stones down into the murky bottom at one time, then heading them off on the hillside. At last they quiet down, and we are glad of it. Midday the 2d July the sheep again had a chance of watering at some sulphur springs. The water smelt very strongly of sulphuretted hydrogen, but was otherwise clear. A bucketful, that stood all night, was in the morning perfectly free from any taint of the gas. There was plenty of swamp grass; the sheep therefore had enough to eat and drink, which enough was of a doubtful quality. Next morning we crossed the flat bottom of the valley, nearly a mile wide, covered with a cake of white and glittering borax; then up a steep hill, and camped not far from Belleville. No water for the sheep at this place, as one may guess when told that it is bought in town at two cents a gallon, for the purposes of washing and drinking.

Belleville is a busy little place. The mill-hands, who superintend the crushing, etc., of the ore, receive six dollars a day. It costs them most of their money to live; in addition, the work is so poisonous, that even with precautions, such as covering their mouth with a sponge while in the mill, few are able to work more than six weeks without falling sick.

It is the 4th of July; the hotels spread a Sunday dinner, and every heart rejoices. It matters little whether American or foreigner, all enjoy the holiday, and are pleased to commemorate in whiskey the 4th of July.

That evening the sheep get water; at night a large bunch skip out and take to the hills, probably with the intention of returning to California. Stock are always restless at first on a drive, and are striving to get back on their home-ranges. It takes the whole day to hunt them, but by evening they are recovered. We are on the edge of the desert, and can hope for no more water for forty miles; this mischance was unlucky, as it took a long day's journey out of half the band without any corresponding gain. Unfortunately we trusted to the men sleeping round the herd to hold them, instead of maintaining the regular sentry watches which I had first instituted. It was a great mistake, and a practice I consented to against my judgment. Nothing keeps stock so quiet as a man moving round them; they lie still and are refreshed, both by sleep and by rest to their legs; which rest they lose if they are fidgety and moving about, hunting food or changing beds, at the instance of fancy or of disturbance.

The San Antonio desert can be crossed in several places, but nowhere is it less than forty miles unless you skirt its upper end, to do which you must go round the sink of the Carson River, which adds to the length of the whole route. It is not a desert in the sense of a sandy waste, for a good deal of bunch-grass grows in little tufts throughout; but water there is none, except in rare tiny springs far up in the hills. Along the road we intend to travel there are some of these small springs which will suffice for the camp and the horses; the sheep must do without till we reach the further side ; for ourselves, too, we must often carry water. What with two barrels, two sheet-iron drums, four milk-cans, and a few smaller vessels, we can start with as much as forty gallons.

In this matter of crossing the desert an ounce of experience is worth a ton of theory. Sheep should be moved quietly, very early in the morning and late at night. On two occasions we let them camp too near the little springs on which we depended ; there would not have been a spoonful apiece for a very small portion of the band, but they were disturbed the whole night, and fighting to reach the water. None of us had crossed a desert before ; and as for useful knowledge, a Mexican, who spoke little English, was the only one among the boys who really understood sheep. He, however, could not take in the situation, and, to my amusement, attacked me in broken English to the effect that I was doing all wrong ; the sheep wanted water every day—at nine and at four. This improvement he suggested in a country where you could not have dipped a pint-mug into water within twenty miles.

Before starting, the trail across the desert was explained by a resident who used to cross some time back, and I hired a young fellow who offered to point out the way, which he said he had in part followed, and in part had been shown. He was a very decent young fellow, but, like the usual citizen, took his money and did not fulfil his share of the bargain. He left us at the only point where there was a doubt ; the road turning sharply to the left, leaving the dry bed of a water-course and mounting on the high ground. He pointed out a detached, fantastic-shaped mountain to the south, as lying in the direction we should follow; whereas the road wound round the foot of another isolated hill—not unlike the one he selected—which stood to the east, and was at the time out of sight. He no doubt wished to earn a few dollars, and, after being out one day, was in a hurry to return home. With this misdirection, which was supported by other details, the sheep were driven several miles south on the wrong road; half a day was lost in looking for the right way, and towards sunset I decided to follow a trail which went east, and seemed to be the most likely direction to follow. This choice proved correct.

Next noon we reached a divide, where we might expect to find a spring from the general description given us. As, however, we were not absolutely certain that we were on the right trail, although travelling in the proper direction, and there was no spring, I had to mount and push on ahead until I should reach something, either settlement or ranch or water. After riding ten miles I came on a few pools, called Black Springs, enough to water half-a-dozen head of stock, that would take their time—not all drink together—and not too proud to quench their thirst with an admixture of sulphur and mud. Satisfied now on the score of the road we were following, I go on another eight miles, taking a right-hand fork, and reach a river-bed lined with willows; there is no water, which has all been absorbed by the irrigation of a farm called Cloverdale. This was disastrous; here was to have been our watering-place at the end of forty miles. I find we must go yet seven miles further.

Having ascertained the lie of the country, and noted

the direction the sheep must follow, I return to the band which has come down to the Black Springs. We try to dam up water sufficient to let the sheep have a drink; but the spring is too weak, the sheep will rush in, and in a minute trample into slush any little water collected. The sheep by this time were suffering a good deal from thirst under a July sun, with dry food, as the grass they found was no longer green; they looked shrivelled, and moved lazily. Those we killed had still a little internal fat. Many of the sheep quite overcame their fear of man, and would push their way into camp attracted by the smell of the water in barrels, and would drink greedily any little quantity of dirty water which had been used for washing. Notwithstanding our many water-cans, with men, horses, and dogs, we had to be very sparing of the fluid, and little could be allowed for washing, except in favor of the cook. If the sheep were camped near the wagon the smell of the water we carried was enough to make those near very restless, and it was all we could do to keep them away from the wagon.

The trouble we had at the Black Springs may be guessed. Although the band was taken a quarter of a mile away they fought us all night long to try and reach the mud-holes; two men could scarcely hold them through the night, and in the morning they break all restraint, and go madly for the spring. Not a hundred, probably, wet their noses; a dozen or so discover where the spring rises, and as a thread of clear water bubbles up here they fight to reach it, trampling over each other, and many sticking fast in the mud. While they were all jostling and struggling we are trying to drag those out of the mud who are not strong enough to save themselves.

With the help of the dogs we drive the herd away and take them forward on their journey; but all the night's restlessness has told upon them; they are tired and travel slowly. The following night, having no water to disturb them, they are quiet.

It was a specially difficult year for driving; there was a very dry winter which reduced all the springs, while the late storms in March had delayed the opening of a road across the Sierras. In ordinary years we would have been crossing the desert more than a fortnight earlier. The sun was extremely hot, and during the noon-day halt the men would build themselves some shade. The Mexican, as an experienced man, taught us all. He would fix on a bush, or a couple which grew not far apart, break away the branches at the foot and pile them on top; over these he would spread his coat; then he would scrape away the upper surface of hot sand till he came to the fresh-feeling ground beneath; a shady little hut was thus soon put together in which the heat was not too unbearable.

On the 12th we passed Cloverdale and made for Indian Gulch. The sheep were now weak and stupid. Next day I hoped they would reach water; they were a long time coming in, and the last of the band were not brought to camp till eight or nine at night. The men were tired, and did not bunch up the sheep as we usually did, so that the watchman could walk easily all round. Partly to avoid hustling the sheep, those that came in last were allowed to be a little spread; in the night a large number again deserted, while we thought they were too tired to move. In the morning, unaware of what had happened, I started out in advance to survey the stream in which at last we would find water and see what dangers awaited us there. Rejoining the herd I found them nooning among bushes and on certain detached rocks at the foot of the hill to which they had been attracted by small patches of shade. They crowded upon one another, trampling the weaker, and sometimes slipping off the rocks; but what struck me at once was the smallness of the band. The men had driven them all the morning and had found out nothing. A man was started off at once on the back trail to follow the strays, which most probably had made for the last springs.

In the evening we tried to bring the band on toward water, which was now not three miles distant. The great heat of the day was over; the high hills threw a good shadow across the pass; but the sheep, oppressed by the long, close, sunny hours, and weak from want of water and food—for they were disinclined to eat, not getting drink—refused to move, apparently preferring to succumb quietly without further ado, or at all events to wait for night. We vainly strove to startle them into a little life; the more unwilling ones we lifted on to their legs—they were very light; others we pushed bodily. We drove them out of the bushes and down from the rocks, and headed them toward the road near which they were lying; it was all in vain. The few we forced to move would slink round a bush and return to lie down with their companions. The dry bed of the river was, I fancy, a little cool and damp; this they would not leave. Luckily a small band of about a hundred came up opportunely. They had taken shelter from the sun under a tree in the morning where they had been left; these travelled more freely as they wished to rejoin the herd,

Seeing them coming on, we took the greatest precaution to prevent their mixing with the main body, and succeeding in keeping them moving steadily along the road; one at a time the others unwillingly rose to follow them. We worked and worked hard in the stifling heat and dust among the bushes, drove bunch by bunch forward so as always to have some sheep on the road and to keep the line unbroken, for it was only so long as the sheep saw others stepping out in front that they would follow.

The longest day comes to an end. Heartily tired of the game, we saw the last sheep well on the road, and then went ahead to ascertain how the band had fared after reaching water. They had come upon it at first in small pools in the bed between steep but not very high banks, which were covered with thick willow and rose-bushes. Following up the pools they met the tail-end of the stream, and in the usual disorderly rout had rushed in. The bottom was pebbly, and the herd mounted the stream so that the water was little disturbed. After a drink they had fallen to on the willows, bushes, and grass, of which there was plenty, and they now looked new beings; they were well filled-out, bright, lively, and not much worse than when they started on the opposite side of the desert. The place was good enough for a day's halt, so I left the band there in the keep of the men, and went back to give what help was necessary in bringing in those which had strayed.

After going more than half-way to the springs I met the boy with some 1500 head; these had, during the night, travelled a distance which had taken them two days to come, and had apparently quenched their thirst at the insignificant spring; anyhow they travelled fairly well. We corralled them that night at the farm, having neither food nor blankets of our own, and as we wanted some rest, and the sheep were squeezing through the bars, I hired two Indians to "heap watch 'em" during the night. We made an early start next day, and that evening the band was all together again. So far as I could judge going along the road no excessive number had died, notwithstanding our misadventures. The principal loss arose from the sheep crowding on each other during the noontide halt, when the weaker ones were trampled down; probably some two hundred sheep had been thus killed in the last two days. Before that time there was no loss, or a very trifling one. The men in camp had taken matters coolly—resting, eating, and sleeping, and not watching the herd. The constant principle of the American help is that work, unless compelled thereto, should be avoided; and all the world over there are no men so intelligently idle as those who have gone West out of Iowa, Missouri, and the neighboring States.

The next few days we had plenty of feed and water; the road went down Reese River to Austin. I got acount on the way down, which I had not been able to do since leaving the stationary camp near the Nevadas; the total of the band had diminished, but how or where was impossible to say. It seemed best to accept past mishaps and take better care in future; our late experience aiding us, we succeeded better, and there were no other losses up to the crossing of the Central Pacific Railway.

It would not be in the least interesting to detail from day to day the recurring duties and recurring annoyances. Nevada is a thirsty land; the little water to be found along the road is being monopolized by individuals, so that stock of all sorts, but more particularly sheep, which are violently disliked by farmers, have a bad time following the emigrant trail. Where there are rivers, the water is taken out for irrigation, and the approaches to the banks fenced. On some of the downstream farms, the people, after the spring freshets, must content themselves with very little water; the upper sluices may be closed once a week, to allow a supply to run down to them, which supply has to be ponded, and is then unfit in a few days for most uses. If the sagebush is cleaned off and the land irrigated, good crops of wheat, barley, and meadow grass are gathered. Vegetables might be grown, but these are troublesome to cultivate, and it is not worth while to make butter or keep poultry. Out of the valley bottoms the country is miserably poor; nothing grows but scrub sage-bush, and a little grass which is found mixed with it. The value of this grass as food for stock is lost, as there is no water for the animals within a far distance, since the enclosures and application of the river. The only towns within reach are mining centres; they help us to replenish provisions, which are, I think, dearer here than in any part of the Northwest of the States.

On the way into Austin we send the sheep round into a canyon for water, which used to flow down to the road, but the rights of the owner below had been bought up, and, the public having no rights, any one driving into town has now fifteen miles to travel without being able to water his horses. Such arrangements are no doubt of a serious inconvenience to a good many people, but no one complains. Hereafter the road may be altered, or if water is brought down the proprietor will be compensated. The herd had to be steered over the hills skirting Austin, which itself lies in a steep ravine; there was no want of food or water near the town, but in one place they came across some poisonous plant of which one or two died.

The day we left, the sheep made a morning march to the top of the pass, from which the road descends into Emigrant Gulch, while I was fully occupied in buying stores, having repairs done to the wagon, the horses shod, etc. Early in the afternoon I ride out alongside the wagon to show the driver the way out of the houses and mills; and when clear of any chance of taking a wrong turn, I push on ahead to the band which are at no great distance, and bid the driver to follow up and give us dinner, for which we are hungry enough after breakfasting at something like five A.M. We wait a long time; it is getting sensibly cooler; the sheep are waking up from nooning, and must soon be taken out along the trail. I ride hastily down the road, and discover that at a very short distance from where I had left the driver, his resolution had failed; the attractions behind were too tempting,—the wheel-marks swung round and went back to town. A little nearer town I met one of the other boys driving the wagon, which he had found with the horses in a livery-stable. My cook and most useful man was last seen in a saloon, trying samples of whiskey and lager beer; quite happy. He turned up next day for his balance of pay, and to recover his clothes, but it suited him to rejoin, of which I was glad. In the mean time the Mexican had officiated as cook, preparing savory dishes, quite undreamed of before; they were good, but so rich that in the three days he practised on us he made me ill. The mutton of these merino sheep seemed to me very good, but the boys preferred both beef and salt pork to it; even the beef was thought the better for a smack of bacon, which it acquired by being fried in grease instead of its own fat.

The food out West in camp is simple and coarse. Nothing but the wonderfully pure air and hard exercise would make it palatable to or digestible by the ordinary mortal. There is, however, no choice—rich or poor, master or man, all sit down to the same provisions, fare alike, and, I may add, enjoy their food. The stock for camp consists of flour, baking-powder, necessary but more or less deleterious, coffee, tea, sugar, and bacon. With a wagon we can afford to carry tins of tomato, green corn, and fruit, a bag of rice and beans, some dried apples and peaches, and a gallon of syrup. These are luxuries; more would be superfluous. The bacon serves the double purpose of supplying grease, in which to fry any meat or fish we can get on the road, or of taking their place when the fresh meat is unobtainable.

This bacon is extremely coarse in flavor and quality; it is often pungent and disagreeable to the untutored palate. It is mostly shipped from Chicago in the shape of large flat sides. The mass of the pigs have been fattened, as the Americans say, behind the steers; that is, in the Western States, where a great deal of Indian corn is grown, it is used to best advantage in feeding cattle; to each steer are added two pigs, who thrive on the undigested particles of corn, and any scraps thrown out of the bins. Those who trust in the chemistry of nature would do well never to inquire closely into the previous food-history of the animals which supply our table.

There is, however, a direct relation between coarse food and coarse flesh; the bacon sent out West is certainly the nastiest of its kind in the world—to adopt the standard of comparison most generally applied by the Americans. A slightly better quality is sold in smaller joints, wrapped in yellow waterproof cloth, and styled breakfast bacon.

The flour is almost always good. It is sold in bags of fifty or one hundred pounds. The bags not in immediate use are great-coated in gunny bags, and put away in the bottom of the wagon, safe from rain and the nibbling of the horses or mules; which at night will often search the wagon for the barley. To make bread the flour is mixed with baking-powder and salt, water is added, and the dough kneaded. It is baked in a Dutch oven, which is a cast-iron pan on three legs; the cover is concave, to hold hot ashes; fire is also put underneath; the baking does not take long if not much water has been used. So soon as one loaf is ready another lump of dough takes its place. This is the most troublesome part of the cook's duty, and he avoids doing it more than once a day. Fresh meat is cut in slices and fried, either in its own fat, or in bacon grease. As to the other mysteries of the kitchen, they need not be revealed here; to discover and to learn them may form part of the project of any one starting out on the prairies. But one mixture is so strange that it must be mentioned, and may perhaps deserve imitation. It has many names; the most harmless is, perhaps, prairie butter. When the meat is fried, if any grease remain in the pan, add flour and water, stir, and mix thoroughly till you produce a frothy batter; spread this on your bread, and, if of my taste—leave it. It is less troublesome to make than butter, which can be the only excuse for its appearance in a ranch; on the road one cannot always help one's-self.

The cook's chief qualities should be cleanliness and despatch; skill comes third; it requires so little, and the boys are so hungry. When the meat is fried and the coffee boiled, a piece of oil-cloth is stretched on the ground, and the necessary number of plates, tin cups, knives, forks, and spoons are set out; the word is given, "Grub pile"; every man washes his face and hands, and, seizing his *couvert*, he helps himself and eats; the cook hands round coffee. After the meat a clean place is scraped in one corner of the plate for syrup, fruit, or pudding, so long as these luxuries hold out. The boys are moderate, except when anything new tickles their palate; then they like to finish it at once. If then the wagon comes within reach they ransack the mess-box, and supplement three hearty meals by an extra lunch. The cook, however, should be a despot, and stand them off; this raid upsets his calculations, and may lead to a second baking. It is the same with whiskey; no self-control will prevent them finishing any given quantity at best speed, thought it is all theirs and might easily last longer. The meat in camp is always better than what is given in small towns, where the butchers and small stock-raisers must combine to clear off all their toughest and most inferior members of the herd. Large owners will never sell locally, but ship every head they sell; this is done to protect their brand, for then no one can pretend to any rightful possession of a cow or a hide with the cattle-owner's mark.

After leaving Austin there was again a dearth of water. We travelled somewhat out of our way to find it, getting what information we could along the road from the few people we passed, whose answers in every case were given with the object of concealing the truth and of misleading us. The road we were on has been for years the main stock-trail, so that one band more or less could hardly matter; yet from some fancy many men would try to turn us away from the right road. I had learnt something of the country before leaving town, but had failed to catch the meaning of various hints and half suggestions, which my informant, more liberal than most, would venture to throw out. He had been an old driver, and knew the tricks of the trade.

A new disagreeable along the emigrant road in this part of the State was the scarcity of feed. By the continual passage of animals—particularly of sheep—the tufts of grass among the sage-bush were either eaten down or trampled out; the brush, too, was dead. It did not stand more than two feet high, but there was not a sign of leaf or the slightest pliancy; the branches were nearly as stiff as cast-iron, and the broken ends hard and pointed. The only way for sheep to get along without having the wool torn from their sides was to stick to the road and the trails along the edges which were partly clear; but here there was no feed, and the only way to get it was to keep the nearest sheep at least a quarter of a mile to one side of the road. Here they could eat, but, as explained above, could not travel, having to work in and out of the bushes. A great deal of wool is torn off by herding sheep through sage-bush; but when you are on the road necessity drives.

CATTLE BRANDS.

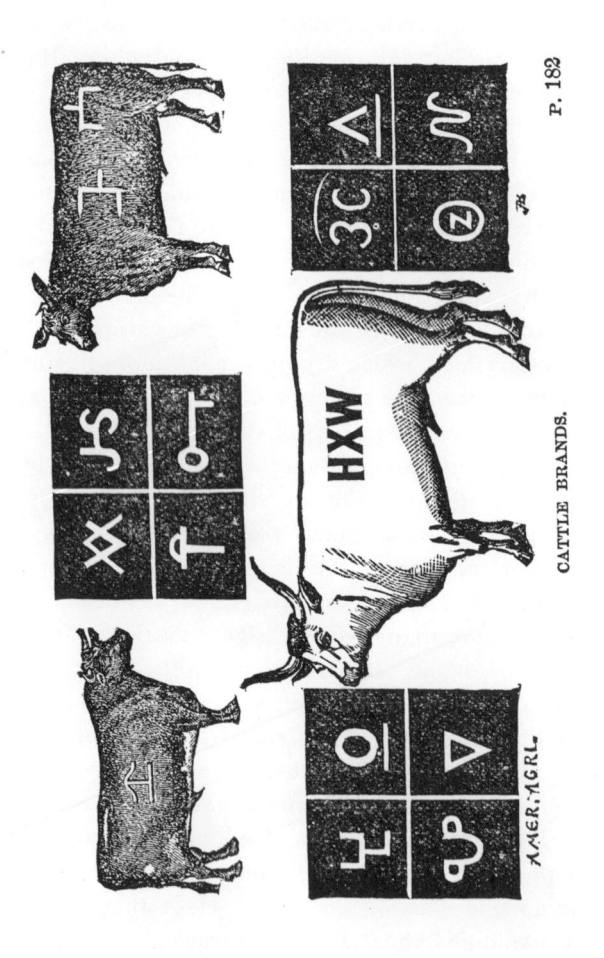

P. 182

Some of the sheep began to lose their wool all over their bodies; this is attributed variously to eating poison-leaves and to fever. These sheep became perfectly naked; a ridge of coarse hair along the back, or a patch of wool like a wig on top of the head, being the last to fall off, gave them a most comical appearance. The new wool began to grow at once, but it would hardly be thick enough to protect the sheep properly through the winter. When a sheep has eaten poison his belly swells; he is unable to travel, and lies down. If we get him as far as camp, as we sometimes did by a ride on the wagon, without any marked signs of distress his head would be stretched outwards and upwards; next morning he would be dead, still looking a healthy sheep, as the swelling of the body was not very marked. A sort of blind staggers was very common, caused, I think, by the heat; you would notice a sheep fall behind the herd, his feet scrambling and working as if independent of his will; he would stumble, run against bushes, trip, fall down. The only remedy for such customers was to catch them and bleed them by cutting a vein in the face below the eye, which almost invariably cured them. They did not lose much blood, and, after walking quietly for half an hour, took up their usual place in the herd. The sheep were also liable to large abscesses on any part of the body; these were cut open, and healed of themselves.

All the time, from after first striking the desert, I was depending entirely on inquiries, and a small scale map, for finding the road. The man on whom I relied for piloting us was worse than useless; he had completely forgotten what parts of the road he had travelled, except a few small details which he jumbled in his memory, and whenever he ventured on a decided opinion he was marvellously out. It was rather a strain having to go ahead and hunt the road, keep an eye on the driving, and arrange for all the wants of men, horses and sheep; for the men would not do anything except walk at the tail of the herd Not one would go ahead to examine a fork in the road, nor ascertain the best place to cross a boggy stream; if there was a check—good, they would stop; if there were two roads, it was not their business to choose. On the other hand, any directions were liable to be received resentfully. It was the boss's business to make all things plain, to make the rough places smooth, to provide plenty of food, and give them plenty of time. How they would have liked to do the work would have been somewhat as follows: A not too early start, drive the sheep along the road regardless of feed, till nooning time, corral them and leave them to watch themselves while the men dine and sleep; at three o'clock drive on those easily found along the road till the camping-ground is reached, corral early, and go to supper; sheep guarded by sleeping round the herd; if they are disturbed and break out, pretend not to hear them; a very little exertion on their part to be made up for by working the dogs. Unless closely looked after, the herders would have starved the sheep by never taking them off the road, or giving them time to feed.

Early in August we crossed the Eureka Railway. Having business in town, I walked a few miles to the ordinary halting-place of the train, Garden Canyon Station, I think, where I found some stacks of wood, and nothing more. As the train was in the habit of being as much as ten hours late, the chance of spending a night among the stacks of wood was only tolerable with the certainty of a good fire. The weather was overcast; heavy showers fell at intervals, which wet my canvas clothes thoroughly. Fortune was better to me than I could be certain about; the train came, and I went into Eureka. Here I bought a small box full of provisions and some barley, but afterwards ascertained that it could not be carried with me in the train. After going backward and forward, and after much hindrance, I find that I can send it in the same train to this empty spot in the prairie by a parcel company, which of course I was obliged to do; they hold the monopoly of carrying by passenger train from the railway company. These little bothers seem queer at a distance, looked at from the midst of more multiplied conveniences; but away in the desert you are glad to find a railway, or store, or anything, and wonder not so much at the price to be paid, and the worries you encounter, but at your extraordinary luck at finding something to pay a price for.

After crossing the Diamond range, we turn northward, and follow down Huntingdon Creek. Before going through the pass, we are assured of water within an easy distance; wishing to spare the horses an unnecessary load up-hill, we only partly filled up our water-kegs. We were certainly very green to trust our informants, for of course there was no water, nor had there been any for a month. The men, being on short commons, vented the usual uncomplimentary remarks on ranchers. In addition, the road was not quite plain; the guide, who had been that way, looked, and said it was not the valley we wanted, which must be over the next range; but I had found out my friend by this time. Next morning, riding ahead to look for water, I came on the ruins of an old post-station, and find a deep well; it takes 120 feet of rope to reach water. I recognize a spot on the map; we decide to noon here, and I go out again to try and make out the road, and, what was equally important, to find water for the sheep.

There was not a sign of habitation: the flat plains brown or gray, and covered with a sage-bush; the hills, some miles off, arid and burning; not even up there can be seen a green patch which might indicate a spring. The soil is loose and gravelly; the water-courses have a coarse, sandy bottom, which would not hold water for five minutes after a shower. It was hopeless to expect to find what we wanted on the plains. The most travelled road I find goes east over the further range; but I judge that the one we ought to follow went down the valley. Making a long detour to take in as much country as possible, I come by and by in sight of a building on the opposite side of the valley, which had previously been hidden by a slight spur. I ride straight for this through the stiff and prickly sage-bush, and find that it is a ranch, only quite untenanted, But, better than all, there is a fine stream of water, which is led down from the mountain. The horse has a good drink; I have a dip; and we then start back again for camp, my doubts and anxieties removed. In camp I point out the road to take the sheep, and explain where they will find water, and, after a lunch, I take the wagon round another way, till we recover the road we had to follow.

Late in the evening the sheep are seen approaching the stream. Naturalists say that the sense of animals is acute in finding water: there must be an exception for sheep. The way they are treated in America makes them much less of a domesticated animal than they are in countries where they are shut all day in fields and closely folded. The argument is generally that natural instincts would be more developed in the former case; but I have known sheep to be exceedingly parched and not be aware of water within a quarter of a mile of a stream. So far as I could judge, they were more guided by the eye, for seeing a green patch of salt grass, or a clump of willows from a good distance away, they would try to reach it. This evening, within a short distance of a considerable stream, it was very difficult to prevent the band from rushing into a mud-spring not fifty feet across, in which a little water stood in holes. But when the leaders did discover the water, and communicated the information by their bleating, it was a sight to see the whole band go off as if possessed of devils—tails up, galloping like deer, and raising as much dust as a regiment of cavalry, baaing their best till they ran into the water, and stopped their mouths in filling themselves. After drinking, the sheep always want to scatter and feed, however late it may be. There was plenty for them, so we gave them an hour and then corralled. But the devils were not driven out, and all night long they made dashes for the scrub to try and go off on foraging expeditions of their own.

So far as I know, the sheep is the greediest of any domesticated animal; he will eat from earliest dawn till the heat of the sun makes him drowsy, when he will lie down. But if at this juncture he gets a drink of water, or sees a change of food, he shakes off sleep and feeds for an extra hour. After nooning he wakes hungry, and while yet half asleep wanders about picking, always fancying something better than what lies in his reach. If he has to move on he wants to stop and grab mouthfuls, as if in a state of starvation, keeping a furtive look on the driver. As the latter approaches the sheep runs on a few paces and then stops and snatches bunches of browze or tufts of grass—swallowing them in haste to snatch again. At this work he continues till his belly is swollen out like a barrel, his back flat, and his legs barely equal to the weight. For a time he checks; but should the herd suddenly cross a patch of some new kind of food, they make a rush, and begin devouring as if they had not seen food that day—snatching, and gobbling, and refusing to move on. Even at night, when half stupid with repletion and sleep, they still try to feed, and finish up on bites of bitter sage-bush, rather than lose the last opportunity. They are very liable to overeat themselves, and when first turned into wheat-stubble, or get among the acorns, they run a great risk of killing themselves. When first put on fresh grass or the young shoots of wild onion they run like mad things; and indeed any morning they will rush past and over good feed, in a wild desire to share in or forestall the leaders' chances.

While travelling through this parched and waterless country, our own condition, as may be guessed, was somewhat grimy. Our outer clothing is made of canvas, which can be bought in every store. The overalls of the herders are generally blue, worn either without under-garments, or over a pair of cloth trousers and red flannel drawers, according to the state of the weather. One or two flannel shirts, usually dark blue, with a turn-down collar, and some ornament, either lacing or buttons, in front; a brown canvas coat lined with flannel; a felt hat with a good wide brim; strong highlows, and a stick. There is hardly ever any difference in the men's working dress from the above; these are the kind universally provided for the Western market, and the woollen goods are worse than inferior. The overalls have to be renewed oftenest. On leaving every town some of the boys would appear in a new blue pair of trousers; a large light-colored patch, sewn into the waistband behind, represents a galloping horse as a trade-mark, and informs all concerned that the wearer is clothed in "Wolf & Neuman's Boss of the Road, with riveted buttons and patent continuous fly." Then come two figures—say 36 and 34—which refer to the size of waist and length of leg. If short and stout, you buy a large man's size, and turn up the bottom of the leg. If, on the contrary, 32 would suit you for waist, you must, not seldom in a country-store, take 40, so as to secure the other dimension. An odd size, however, leads to tailoring in camp, which is an unprofitable employment; most men, therefore, start with at least one extra pair of overalls to fit. The patch is left either from idleness or as a memorandum of one's measurements.

For the rough and dusty work of driving, whether on horseback or on foot, these canvas suits are the most efficient; they turn wind and dirt, and can be washed. Where you have to follow stock in a cloud of dust, and have the ground as your only seat, woollen outer garments would be objectionable. In cold weather, therefore, you put the canvas overalls and coat over the woollen ordinary clothes; they make a great difference, and help immensely in keeping you warm. In July and August, of course, there was no cold to contend with by day even at high altitudes. Whenever sufficient water could be found, and a little leisure secured, it was a great achievement to have a bath. Dust is so penetrating, that the least said about one's condition is best said. It was a great consolation that it was clean dirt, for after having washed thoroughly, a quarter of an hour at the tail of the herd would blacken you as before. In truth the occupation is so laborious, the hours so long, and the attention must be so unremitting, that a bath is often out of the question, even when the quantity of water is to be found, for those who have to do the work. The middle of the day is the only time available, as the drives are arranged for the stock to water at that time. The wagon generally gets ahead in order to fill up kegs before the stock come in and trample the stream into mud, which takes but a few minutes after they arrive; the men come in at the tail: to find clean water they must go off half a mile. To bathe in the evening long after sunset, or in the early morning when you should have finished breakfast by sunrise, is out of the question: first, you are too tired; secondly, it is too cold even in summer among the hills. and—twentiethly, you are very seldom camped on water, If by luck you find yourself near a deep slowly-flowing stream, in which the water is warmed a little by the sun, it is a festive day. There is generally feed on the banks;

the sheep, which prefer slightly warm water to a bright cold rivulet, are content to stop round; you then can go in for real luxury, bathe, change and wash the clothes you take off. In the evening you are again as before, the bath but a memory.

The natural result of these circumstances is that the boys seldom look to ablution beyond washing their faces and hands. They are careful in this. Barring dust, it is a clean country, and there is plenty of fresh air. Dirty men abound, and at least one is to be found in every outfit; but his habits are very freely criticised, and sharing of bedding or clothes is carefully avoided; it is fate that he should be there; you must put up with him, at least for a time.

The bedding consists only of blankets or quilted counterpanes; your pillow is a bag stuffed with your spare clothing. If possible, the whole should be contained in a sheet of extra stout canvas, sufficiently long to be spread underneath you, and when brought over to cover you fully; the width must allow of a wide margin being tucked under the sides; about fifteen feet by seven answers well. At night you spread your bed on the ground, and if the sides are properly tucked in, should it come on to rain, you draw the upper fly over your head and lie snug; the canvas is fairly waterproof. In the morning you turn the edges inward on top, roll up the bed, strap or tie it tightly. The canvas keeps the bedding clean and dry, protecting it against dust and objectionable emigrants, who find themselves crowded in other blankets. Usually the boys sleep in pairs, which increases their resources and saves weight; the bedding is the bulkiest part of the load in the wagon. Your night toilet consists in taking off your coat and boots; the coat you may imagine a pillow, your boots must be tucked away safely to keep them dry, and beyond the reach of cayotes, who will steal into camp at night and carry off anything made of leather; without your boots you would be in a very poor fix on the prairies.

As in all elevated countries, the difference of temperature during the day in the sun from that at night is very great. Although you may work in a single flannel shirt, it is proper to have plenty of blankets for your bed. It is the cook's duty, after fetching camp in the evening, having unhitched the team, to tumble all the beds out of the wagon on to the ground. Each boy at night carries his bed to a spot he likes and there unrolls it; he is limited to some definite direction, from which he is supposed to assist in guarding the sheep. It is not always a search which ends successfully. When you start after supper in the dark, carrying a heavy load of bedding with the purpose of making your bed, the ground may be sloping, and thickly covered with sage-bush; there are hollows, and sidling places, and stones, but there is no level spot even six feet by three. You are a little out of breath with the weight on your shoulders; it leans against your head, which you hold sideways; you cannot see clearly, and stumble up against bushes or trip over stumps in the dark; you drop your bed carelessly with a flop, and —up jump the sheep. Having jumped up they begin to stray from their bed-ground in search of feed. Your first business must now be to drive them back and watch them till they lie down and are still again; you may then return to your bed, and after spreading it out as much as can be done in a narrow space between the bushes, you pull off your boots and creep inside the blankets. But where is comfort?—a root-stump is under the very middle of your bed, invisible to your eyes in the dust, but prominent to your present feelings. It is, however, a very aggressive stump that makes you shift your quarters. You are far too tired to mind a little bullying; if by any means of bending yourself into a C or S curve you can avoid the knotty point, it is good enough, you and the stump need not fall out; anyhow you are not going to move, and will, you hope, sleep soundly.

Granted that your expectations are accomplished; suppose the sheep have fed and drunk well during the day, and therefore are not inclined to move that night; say that there is no wind-storm to disturb you, and the plaintive cayote is dumb, the hours pass too quickly; you wake in the dull-gray light of break of day, a little flame is flickering in camp, the cook's voice shouts "Roll out;" you jump up, but before you have time to dress and pack your bed it is "Breakfast!" Your carry your bedding to the wagon and dump it down somewhere handy. Having washed your face and hands, you take a place near the fire; somebody throws on a bush to make a blaze, and you eat a hearty meal of fried meat, bread, and coffee. Long before you are ready the sheep are on the move, and break up their camp; if they travel in the right direction you can let them go, but if wandering, one man must start at once and take charge; the remainder of the boys finish breakfast, fill their canteens with water, grasp their sticks, and follow the herd. The cook is left in solitary possession; he has to wash up, reload the wagon, catch, feed, and water the team, and then follow the trail of the sheep, and be up again in time to cook dinner.

Every night does not pass comfortably. After a rough day's work, or at the end of a longer journey than usual, the sheep, not having crossed good pasturage, or having been driven too quickly for leisure to feed so much as they want, are restless and hungry. When first corralled they may be feeling a little tired, and will lie still a couple of hours, just enough to lull you into an idea of security. You are no sooner in bed than they begin to break out and wander away in search of food. A dozen well-known brutes are the leaders; drive them in at one point, they will thread their way through those lying down to another point of the bed-ground, and try the old manœuvre. It is useless to let them have their way, thinking they will satisfy themselves hanging round the herd and then return. They have not gone out five minutes when a hundred others are up. If feeding is to be allowed, they also want a share. To avoid close quarters each lot creeps out a little further; nibble and move, and nibble again: this is only the beginning. You must check it, and, deferring sleep, commence to walk round and round. At every point they leave the herd you must drive them back. If no one notices them at the first it does not take ten minutes for the whole flock to be moving. A suspicion comes to you in your sleep; you wake and hear the peculiar creeping sound from thousands of tiny hoofs; the bells are making a noise—you need hardly look at the bed-ground—the fact is already too certain. You must jump out of bed sharp, pull on your boots, seize your coat, and go after them at once.

When the sheep find themselves discovered they stop, and if you shout at them they turn and come back; even those a good distance off will hesitate when those behind are leaving them. They stop, look, and soon begin to run inwards; those furthest off are even a little scared, and come galloping in, charging the bushes, and tearing off wool. You must go out to the utmost point they reached to be certain none are left; walk all round in a big circle, and then come back and bunch up the herd, watching them till they lie down. This game you may have to repeat three and four times in a night. As the hours advance and the temperature falls you stir less readily, and hope by shouting to turn the wanderers. This will not pay more than once or twice; you find you must get up or lose the sheep. Many a time it seemed better to lose the sheep. In the morning it is as well to make a wide sweep, and assure yourself by the tracks that none got away. You can also count the bells, and, if possible, the black sheep. On moonlight nights the sheep are bolder, and are more likely to stray than on dark nights. It is after all least trouble to watch them regularly, though many experienced drivers have told me that they had always been satisfied with the men lying round the band. Yearlings are the most difficult class of sheep to deal with on the drive, and Merinos are said to be particularly "mean." That was certainly an opinion with which I came to agree.

The trouble varies inversely with the quantity of food.

With plenty to eat, and not too far to travel, the sheep will lie so still that at sunrise they are still chewing the cud, and have almost to be kicked off their bed-ground. When hungry let the sheep have a quarter of an hour's start, they are out of sight. Going with their own inclination, they cover a great distance in a few hours of night; but along the road they take some time to understand the drill; and you must not venture to require more than an average of nine miles. They do come to understand that they have to follow the direction of a road, and it is interesting to see them sometimes, when brought back from feeding to the main trail: the leaders check, look along the road, look back, and, finally making up their mind, start off in the proper direction.

The cayotes sometimes worried us. If one came near the band the whole of the sheep on that side would start to their feet; this takes but a second. They are wonderfully quick and unanimous in doing anything disagreeable. That part of the bed-ground is deserted; the disturbed sheep pass round the remainder of the herd to the further side; another batch is then perhaps frightened; they move away. It does not take long for the whole herd to be on foot. The good herder will now do his duty, but the hireling sleepeth. He has got up once in the early part of the night; it is colder now. If he has a dog he will send him to round-up the band, but the dog soon tires of being constantly sent out. When his master does not accompany him the dog shirks and sneaks back to the bed without having driven in the sheep. If he is threatened or punished he simply runs away and lies down near the wagon or in the bush till the morning. As a work of supererogation the dogs used to take a great delight in chasing the cayote. If one howled near camp all the dogs would start in pursuit; they ran and barked till they tired, and then returned. They never tackled the cayote if they caught him. This play was disturbing both to ourselves and to the herd. No good came of it, so I had the dogs tied. The cayotes never did attack the sheep at night. One or two were killed by them on the road; it was done by daylight.

Well-bred and well-broken dogs fetch a good price, if you can hold them till you find a purchaser who is really in want of such an animal. The day-dream of a herder is to get a dog that will watch the sheep at night; for even to wake and halloa a few times makes a bad night, and no one need envy the man whose fate compels him to walk half-chilled round and round a lot of fractious, pig-headed sheep; to find the same brutes leading off again and again, bunches watching him, standing still as statues in his presence, but stealing out from the corner on which he has just turned his back. If he sits down on a stone for ten minutes the whole of the work seems to have to be done over again. He comes on a band that he has headed back already half-a-dozen times. They wait till the last minute, and trot into the herd just a yard in front of him; so soon as he is past they walk out. You must take it slowly—impatience would do more harm than good; for the sheep you drove in with a rush would startle ten times their number among those which perhaps had been lying down; they then pack and squeeze on the centre—heads inward, tails outward. The chief culprits have knowingly secured forward places quite of reach. The lot cannot remain so, and to lie down must open out. You have to leave them.

Quietness, patience, and persistency—these are the cardinal qualities; keep on turning them back until they are all lying down; you may then go to bed. But in the first instance choose your bed-ground; have room enough for the herd to lie down without crowding; they will lie all the quieter for a little elbow-room. Any place does not suit a sheep's idea of comfort. If a big wether sees a smaller sheep in a spot he fancies, he will touch him with his fore-foot as a signal to clear out; if the sheep will not take the hint the big one will butt him out. On several occasions when the sheep had been particularly well-fed, and were proportionately content, they spread out their ranks till in the morning they were seen lying all round the men's beds, and in the closest proximity thereto; but at these times they did not care to feed at night. Properly handled sheep like nothing better than to carry out their *rôle*, which is to grow wool and to grow fat; it is for the men to help them to do so.

Good dogs are of enormous assistance on a drive. They are scarce in California in the early summer, when every band going to the hills wants two or three dogs. Some owners pretend they would rather be without dogs. It is possible that in driving fat sheep in the plains the men would work the herd more quietly than the average dog; but they are a necessity where the ground is rough and covered with bush, and if the sheep, attracted by some new food they are fond of, are liable to scatter, dogs get them in more quickly than any man can o, and by turning those heading in a wrong direction at once save time and save the sheep an unnecessary journey. Sheep, too, will mind a single dog, when they would not be controlled by several men. The watch the latter, and dodge them so soon as their attention is engaged elsewhere. A dog who has nipped them once or twice instils a wholesome fear, and for him they will turn at once. In bad hands a dog is liable to be rough. A lazy man will spoil his dog by overworking him; the dog learns bad tricks, and saves himself by cutting across little bunches instead of going outside all.

The relations between the sheep and the dogs are amusingly various. One of the two I bought was completely master of the situation. He had only to show himself when order was restored; the other was an older dog, of a reddish-brown color, "a yaller dogue," as the Americans say. Originally broken in as a sheep-dog, he had in later years been used in hunting bears, and for various farm purposes; he was strangely gentle notwithstanding; the sheep all found out his disposition, and were quite friendly, allowing him to walk close alongside when lying down, with other liberties. This mutual good-will was a disadvantage; as a straggling sheep, being admonished from behind, would turn and face the yellow dog, when a comic fight would ensue about nothing at all, and in which neither attempted to hurt the other. At other times we have all seen him go up to a sheep, who would look at him inquiringly and not move; the dog would lick the sheep's mouth. Both dogs had little tricks, in which they had to be checked, as furthering their own ends without regard to our intentions; in fact, every evening they were so anxious to conclude the day's work, that, if allowed, they would have driven the sheep alone into camp at a great pace. At night they hated to work as much as any of us, and would hide carefully if called upon too often to assist.

An amusement of theirs, of which they had to be broken, was hunting jack-rabbits; these, latterly, were very plentiful near water. The long line of sheep traversing the country would often put up quite a number of these rabbits, who could not always readily get out of the way; they would run ahead and sit up, and look and run again; some, finding themselves caught in the midst of the flock, lost their heads, and rushed wildly about among the sheep, giving the herders a chance of knocking one over with a well-directed stick. Chipmunks might also occasionally divert the attention of the dogs. They seem to resemble exactly the small striped squirrel in India; the latter live in trees, the former on the ground. Ground-squirrels I saw in Washington Territory; they are excessively destructive to the crops of the first settlers, but as the woods are cleared they are said to go back. Prairie-dogs were not common in Nevada. There were plenty on the prairies we crossed in Wyoming and Montana. Their bark is more like a chirrup; they are pretty, fat little beasts, seen sitting upon the mounds which surround the mouths of their burrows; they eat the grass very close round their village, but are otherwise harmless. On the other hand, as they are of no use to us as food, we naturally slight them.

A very early remark an Anglo-Indian makes in the West of America, is the resemblance of nature and of animal life to the types he knows in India, and the dissimilarity to European specimens. The Rocky Mountains remind one strongly of the Himalayas; a resemblance which might lead one to suppose that gold may yet be found in the latter mountains, if searched for by experienced prospectors: geologists are of no use at this trade; while the ordinary uninstructed individual would build his hut against a quartz ledge, and not suspect its existence. People who have explored the Himalayas looked at the scenery, or were searching for game. Judging from American experience, mines are found in places as wild and difficult of access as any happy hunting-grounds of ibex and markhor. As for the animals, I will only mention a few: the cayote, in appearance and language, is a jackal; the American antelope, in form and habits, resembles the black buck; the elk corresponds with the bara singha; the buffalo with the Central India bison. As to the surface of the country, the wide plains and lofty plateaus have nothing to compare with near us, but can readily be matched in Asia. The light soil, fertile only under irrigation, and when cut up by water developing perpendicular-sided ravines, which cut back and back, plains covered with saline efflorescence—these can all be seen in Colorado, just the same as in India. There is, however, one great difference—the sun. America develops our race, India kills it.

In going down Huntington Creek we had left the main travelled road, and were making our way along the foot of a range in the direction of Elcho. There are some charming valleys embosomed in the mountains, with plenty of water for irrigation, and many farms. Only one herd had travelled this road during the season; feed and water were in good quantity, but wire fences too common. The main valley itself was as brown and arid as the country generally we had come through. I lost a horse by his own clumsiness; he was accustomed to be hobbled and turned loose to feed; in crossing a small stream he must have fallen backward, and was drowned in about a foot of water. I had some difficulty in replacing him; the opportunity of a forced purchase raised prices against me. I was offered any old screw, and asked to pay out of all proportion to his age and infirmities.

But I would not offer the horse-dealer as the type of the business-man any more in the West than all over the earth. Tell him what you want, and see what he brings you. His object is not to suit his customers, but to get rid of some objectionable quadruped in his own stables. Tell him you would like a horse fifteen hands, bay or brown, not more than eight years old, and in America you would add his weight—say 1200 pounds. He has quite a number to suit you; if you will call again to-morrow morning he will have them driven into the corral. Judge of your surprise!—you see some half-dozen miserable-looking beasts; two are mere ponies; one lame, the best one belongs to a pair; the last, which he strongly insists is just the horse to suit you, is gray, scraggy, over sixteen hands, out of condition, and never likely to see eighteen years again. His friends helping, he would wish to worry you into taking this venerable relict off his hands.

By the end of the month we had crossed the Central

Pacific Railway, and rejoined the main trail. The country was still much the same; feed scarce along the trail, and watering-places a good distance apart; but the country was higher and more mountainous, and at times picturesque. The road lay just across the corner of Idaho and Utah, and passes a spot marked on most maps as City of Rocks. Many persons have fancied that there is a town here, but the name has been given on account of some fantastic masses of rock which stand round a valley; they are not very remarkable, and are not to be compared with the stones near Manitou, Colorado, called the Garden of the Gods. The weather in September was growing very chill, and over the high ground we had often frost. We had to be constantly on the watch in case we should find ourselves camped in a place without firewood; this was troublesome, as a good fire was an imperative necessity while eating supper, both for warmth and light. We often carried wood along in the wagon, but that hardly sufficed for camp-fires, which to be of any account must be fed with prodigality.

For the only time on the journey we were one night disturbed by a bear, who was accustomed to travel up the road to a small spring of water. The sheep, though not attacked, were greatly terrified, and went off bodily up the side of a hill; we brought them back, and with a little watching they settled down quietly for the rest of the night. Bears in some places are very bold, and will climb over the corral fence and kill several sheep. But I will speak of them as I found them; they never did us harm. I tried, in an idle fashion, to shoot the one mentioned above. The injury was not completed; my intentions were evil; the bears, on the contrary, behaved well to us. The most zealous sportsman would hardly care to roam in the dark with a cold breeze blowing up a canyon on the off-chance of a shot.

The only possibility of keeping warm at night was, before getting into bed, to put on an overcoat, and turn in all standing; minus boots most people would do it; but a good pair of loose boots are no detriment in bed, particularly if there are likely to be calls. They don't hurt the bed, and keeping them on saves time.

As we get into Idaho there is a marked improvement in the country. Grass and water are more plentiful. There are cottonwood and birch trees not only along the streams, but in fringes on the hillside; and wherever a hollow has retained the snow after its general disappearance from the ridges of the hills and from open spots, the later moisture has encouraged the growth of everything green. But the autumn is decidedly fading into early winter. The higher ranges have once or twice been capped with snow; the leaves are changing from green into more lively colors; the sun in the middle of the day even is occasionally feeble, having probably overworked itself in scorching us through the summer. It was high time to consider where the sheep should be wintered. The choice lay between taking them south to the country which borders the Salt Lake, or to push on either to Green River, or to the Laramie Plains. The Green River country was, however, said to have been overstocked for many years, and, although ranges may still be found, good ones are scarce, and without plenty of feed a band of sheep, more particularly one which has travelled up from a warmer climate, would have a poor chance in the extreme cold of these parts.

The Laramie Plains are a portion of the highest tableland between the oceans; although subject to as bitter cold as anywhere in the northwest of the States, its exposed position, liable to be swept by strong winds, enables stock to live, for the reason that, the snow being blown off, the herbage is laid bare. This is the case in ordinary winters; animals which start healthy and in good condition pull through on these plains fairly well; but in every season there are severe snow-storms and piercing winds, during which it is impossible to take out sheep, and when cattle and horses cannot do better for themselves than turn tail to the blast and drift slowly before the storm. The chinook, which is a warm wind, blows at times and melts the snow; but the greatest danger to all stock is when such a partial thaw is followed by sharp frost. The surface of the snow is then ice-bound, and it is impossible for any animal to care for itself. To meet these cases a sufficient quantity of hay must be provided for the sheep; if not, the chance may be that the whole herd are starved and frozen to death. Even with hay in hand it is not always a good plan to feed it to the herd, for they will not in future take the trouble to hunt for themselves, but idle all the day and wait for the hay in the evening, a process exasperating to the easiest-tempered herder, but all in a piece with the general behavior of sheep.

The climate of the country lying to the south of and surrounding the Salt Lake is much milder than that of the nearest portions of Idaho or Wyoming; the snow does not lie deeply, and the plains, besides grass, bear the white sage, which is very nutritious. The latter, after it has been nipped by frost, is apparently much relished by all stock. A light fall of snow here is an advantage, as it enables the herds to push out into the plains, which are waterless; the sheep can eat snow, and the herders melt it.

On these trips the herders live in a small canvas house, which is built on to the wagon; in this there is a stove; the bed is on a low shelf across the hinder end, the entrance is on one side. With the traps and supplies of a couple of men, two horses are all that are required; the wagon does not move every day, and the journeys on occasions are short. To men who are not averse to a solitary life, and do not fear rough times and exposure, this wintering with sheep may be tolerable. A man who understands the work, and can be trusted to do it, should always be able to secure something better than good wages. There are plenty of men in Utah who, having saved money, would like to invest it in a band of sheep. The sheep, to live, must travel summer and winter. It is impossible for a man resident in a town, and with a business, to see after his sheep in person; he therefore has to look round either for a herder to manage for him, or a joint-owner to share in the speculation. The current expenses are not heavy; two men can through the year easily drive two or three thousand sheep, with a little help at lambing-time. The returns from wool and increase are not exaggerated at twenty-five per cent. As the profit with sheep much more than with other stock depends on the care and success of the men in charge, the man who knows has a power which in some cases transfers the flock from the owner's hands into his own in three or four years. The alternative, therefore, to the proprietor who cannot accompany his own herd often lies between seeing his property destroyed through ignorance or transferred through cuteness. There is, therefore, a good opening for any man thoroughly versed in sheep business to make his way in Utah.

To make satisfactory arrangements after reaching Wyoming to winter the sheep properly required more time than the season promised. If a purchaser were found readily they might be sold to better advantage than they could be in Utah, but the readiest plan was either to sell on the spot or to drive south. To one not trained to the profession it will not be difficult to fancy that the outlook of a long winter, in addition to the summer spent in constant shiftings, dirt, and annoyances, would be somewhat dreary. Whatever the profit, it might be earned too expensively.

As the road would soon cross a railway, I left the herd to follow slowly, and pushed on. The railway-station where the trail crosses is called Arimo; the town is called Oneida, and is marked in the railway guide-books as having, I think, 600 inhabitants. It was hardly a matter of surprise to find that the two places together consisted of a small station-building, a store, a couple of farms, and the wagons of an engineering train. The inhabitants, excluding railway workmen, who were only there temporarily, might number a dozen. It is one of those places the ordinary traveller looks out upon from a comfortably warm Pullman car, and vainly tries to imagine what inducement there is to persons to live in such a waste. The few dwellings stand with the air of frozen-out laborers; the surroundings are dwarfed sage-bushes, almost to the walls of the houses; the men he sees wander listlessly along the railway embankment, unoccupied, serious, woful. As they walk down the train they pry into the windows, as if longing for a face they may recognize. The post-master comes dragging a mighty leather bag, which is locked but empty, and changes it for another equally empty. A word is spoken to the conductor, the bell rings, the train leaves the sad-looking colony; the men in a disheartened manner walk back to the store. Half the excitement of the day is gone.

But any station is good enough where you can get on to the cars and rejoin civilization, though the process at twelve miles an hour seems unnecessarily prolonged. The railroad passes through Cash Valley, which has been made a garden by the industry of the Mormons, and joins the Central Pacific at Ogden; thence another line runs to Salt Lake City.

It is impossible to mention either Utah or Salt Lake City without adding one's impressions with regard to the Mormons and their religion. Much has been done lately, by persons who have studied both, to put before the English public a less prejudiced version than that which has heretofore reached us. I had not the time nor the opportunities to study the matter, and my slight personal experience was from contact with the farmers of a few colonies which were passed on the road. The opinion formed was entirely favorable to the religionists; they are industrious, thrifty, quiet, and fair-dealing. It was a great good chance when we came among them, as we were able at once to obtain dairy produce and many vegetables which were not procurable among the usual American farmers. They have the credit of being hospitable and kindly natured to travellers or persons living or working among them. They seem to try more to live up to the doctrines common to all Christian religions, and are more earnest than the followers of most other sects; they are great observers of the Sabbath. All dispassionate people like them; it is in Salt Lake City that they are mostly abused. One paper of that town in particular seems to hold a brief for persistent and rabid accusations; no sin is too gross to lay to their charge, no theory too far-fetched on which to found a plausible reason for their persecution. The notion even of a coming degenerate and brutalized race was seriously debated in an article, which race would be found in the offspring of mothers raging with all the passions of love curdled into hatred, and writhing all day under the injustice and tyranny of polygamous men. The Supreme Government was called upon to interfere in the interests of the future of the State, which would be marred by the aforesaid race of monsters, who by the laws of heredity would naturally turn to vice and crime.

I had the chance of seeing a very large number of the Mormons, as Salt Lake City was full of people who had come in for Conference; this is both an occasion for business and pleasure. It was impossible to notice any signs of the grief and oppression which is said to mark the Mormon's wife; the character of the faces were exactly the same as seen in other Western towns. The younger and American-born women were bright enough, somewhat showily dressed, and looked quite equal to the care of themselves and the safety of their interests without other assistance; the older women had the hard, grim faces which we associate with Puritanism and that kind of fanaticism which suppresses all amusement.

This dull, expressionless sort of face is universal out West, and is probably traceable to the cheerless lives, hard work, and grinding poverty experienced in younger days. The men looked neither to be the better nor the worse for the special advantages they enjoy. The fact is, polygamy to the outsider stands prominently as an ugly feature, on which the whole of Mormonism is based; and that it depends on this foundation alone for its existence is a hasty conclusion of the uninitiated. I do not pretend to know; whereas, so far as I could judge, Mormonism is maintained mostly by the Church government. So long as the mass of this sect allow themselves to be used for the purposes of extending their religion, and are managed by clever men, the sect will prosper. The power given into the hand of their leaders has so far been wielded with a good deal of business-like capacity, and to the benefit both of the religion and the temporal prosperity of the community; and if the leaders had not continually to fight their opponents, no doubt the funds raised for the Church would be used to better purposes than in defending their peculiar institution.

After all, polygamy is the only tangible ground on which the State can interfere with the Mormons. This practice is so contrary to deep-rooted because long-accepted convictions that it cannot last, and the idea of a sect numbering several hundred thousand being able to oppose the declared intentions of fifty millions is absurd; plurality of wives must and will die out. But where the Mormons make most enemies is by the assumption of their being a separate people. It is not reasonable to talk of the Mormon people; they are recruited from the same races as supply the bulk of the emigrants who surround them; but, as is well known, they like to talk of themselves as a race specially chosen by Divine inspiration, and, in the usual biblical jargon which most sects adopt in their inception, they tell the world that they are set apart for the regeneration of mankind. One of their fancies is, that America is to be harried for her sins, and to be saved by the remnant of righteous people in Utah. This talk is harmless enough, but is hateful to the good American, who allows of no divergence from American views. This segregation of the Mormons, in conjunction with their Church government, is made the ground of accusations of disloyalty, and when such an open defiance of the States law is insisted upon as claiming that polygamy is necessary to the practice of their religion, a good handle is offered to their enemies. They are called bad citizens, than which no worse can be said for them from an American point of view.

I asked a few persons if the tithe paid to the Church was not a heavy tax; these did not consider it so, and said it was given voluntarily. On the other matter curiosity would have been allied to impertinence. Among the poorer class a second wife is, I think, now rare. That the second wife is, in many cases at least, assented to by the first is apparently true enough. I was curious to discover how this consent could be brought about in any single instance, arguing of course from the premise that part of such a valuable property as a husband was not equal to the whole. The only partially satisfactory conclusion arrived at was the stress of in-door work. Men for out-door employment are always to be had, or nearly always, but a woman for household labor is scarce and unusual; the whole of the cleaning and cooking falls on the wife. When, therefore, a man thrives, has a family, and hires half-a-dozen hands, the wife's share of work is considerably increased; at the same time, she may have grown stiff in the joints, and feel that a certain indulgence of ease is both natural at her age and attainable under the circumstaccs. What better plan, where the power to add is part of the articles, than to secure a third partner in the firm, which insures solidarity, whereas a hireling help, if obtainable, could not be entirely trusted, and Madam No. 1 would find her hands still full in assisting and watching a paid servant. I will not say that this was the motive in any particular instance, but it may be inferred that some such process might have led to a dual control of household matters. The explanation is reasonable in itself, and, others failing, may deserve consideration.

The Mormons pretend that they will not resist persecution by any means, except by prayer to God to soften the hearts of their adversaries, and that they will trust entirely to Him to protect them and to establish their religion. It would be more practical if they would look ahead, see what is the tendency of their religious practices, and, if possible, not run their heads against a wall; they wont hurt the wall. Probably owing to the fact that the Mormons are being roughly used, the sympathy of an entire stranger to American ways is stirred in their favor. A short residence in Salt Lake City, where the feeling against them is particularly hostile, and is violently expressed, would naturally rouse an unprejudiced person in their defence, and possibly at the same time make one tolerant of all their fads, which would otherwise deserve no serious consideration. The fire on the altar of Mormonism is burning down; if the politicians will leave it alone it will die out quietly. Nothing can be brighter or lovelier than the autumn tints on the hillsides which overlook Salt Lake City. The town itself is attractive, but falls far behind both Denver and San Francisco in size or importance. It is not on the main line, nor is the country around sufficiently fertile for a high state of culture, notwithstanding the extravagant praises which it has been customary to bestow on the efforts of the first emigrants into the valley. There are valuable mines to the east up in the mountains, and there is some talk of a future boom in mines in the country south of the great lake. The latter, as usual, are some of the richest mines in the world ; but one hears the statement so frequently in diverse places that it fails to impress, particularly when accompanied by a readiness to supply every one with some of this invaluable property. The only thing the miners are waiting for is the extension of the railway, and this, as usual, has been located in two places. The whole region in that direction is, from all accounts, barren. What advantage a line of rails could obtain by crossing it is less than doubtful ; but some railroad is sure to be found to undertake it, if only to forestall a rival ; possibly an extension of the Denver and Rio would meet the case. Nothing could be less paying than the country which is traversed by that line between Salt Lake City and Pueblo. It is advertised as the scenic line, and a short length does pass through some most Titanic rock scenery—an extension through waterless deserts may afford a happy contrast. A railway is often followed by a feverish spurt of speculation, and one or two mining towns will probably spring up. These will revel in the glories of numberless saloons for a season, and in a short time after fall flat as many other places have done ; for instance, Aurora. But how these Western railways, running through a poor country, are to pay the different companies who finance them, stock them, issue first preferences on them, and water their shares, is a branch of business not given to every fellow to understand.

A very short stay in Salt Lake City satisfies most persons. It certainly may be called a pretty town—the trees and gardens having a good effect ; but how long would the latter be retained when the land becomes valuable? Still at present worse places can easily be found, and when the burning question is settled the town will probably take a fresh start.

This narration will, I think, give a truthful impression of the manner of life which must be followed on the trail in the Pacific States and territories. It is not everywhere so dry and so dusty as in Nevada ; but with certain allowances for the pleasanter aspects of affairs in journeying through a better grassed and better watered country, any one can fancy for himself how far he is likely to appreciate the life. There may be difficulties special to that portion of the territories lying further north, owing to heavier timber and bush into which sheep might stray, and to the greater cold and deeper snow which prevail through a longer winter. But wherever it is followed the business of driving or looking after sheep is rude and tiresome. The daily companionship of less educated men is wearisome ; the out-door life is healthy and exhilarating ; the roughing does not show too disagreeably.

Young men who are fitted out with good spirits and manliness have nothing to dread. America is a land of hope, though often of hope deferred. It is well to go and see it for yourself.

One thing is perhaps surprising if you think of it; that is, why young men are satisfied to do the work of farm laborers in the West, with all the added discomfort of coarse, bad fare, separation from their friends and associates, and a complete loss of mental culture. Why are they content to plough and cart manure in Iowa; to herd sheep in California, or cattle in the Rocky Mountains? Are similar occupations less derogatory because carried out at a distance from our homes? Are they derogatory at all? For choice, labor on an English farm would be lighter and pleasanter; the food and shelter at the smallest farm-house much better than on the wild prairies at the ranch of the most wealthy cattle-owner. In the one case there are supposed to be chances; are there none in the old country? We hear constantly of working-men and mechanics who have tried America returning to England to better themselves. It is just possible that a young man who would himself work through the drudgery of farming in England, and live down to a very modest scale, would take more pleasure out of life, at the same time escape much of that side of frontier existence which, when I heard it called "beastly," I could hardly feel the term misapplied. This is not perhaps the happiest way to conclude my story of prairie experiences. As opinions one must say what one thinks, and as facts not more than one knows, if within those limits none have a right to accuse you.